霍德盖茨＋冯
Report / 2005

Acknowledgements

This publication has been made possible with the help and cooperation of many individuals and institutions. Grateful acknowledgement is made to Hodgetts+Fung, for its inspiring work and for its kind support in the preparation of this book on Hodgetts+Fung for the AADCU Book Series of Contemporary Architects Studio Report In The United States.

Photo Credits:
All photographs are by Hodgetts+Fung unless otherwise noted below.

Tom Bonner: 54, 55; Benny Chan: 114; Sally Painter: 60 (lower), 62 (center), 64 (center, right); Marvin Rand: 89, 93 (upper), 94 (left), 95, 103 (lower right), 105, 106-107; Alex Vertikoff: 35 (center), 36.

©Hodgetts + Fung
©All rights reserved. No part of this publication may be reproduced, stored in a retrieval system or transmitted in any form or by means, electronic, mechanical, photocopying, recording or otherwise, without the permission of AADCU.

Office of Publications:
United Asia Art & Design Cooperation
www.aadcu.org
info@aadcu.org

Project Director:
Bruce Q. Lan

Coordinator:
Robin Luo

Edited and published by:
Beijing Office, United Asia Art & Design Cooperation
bj-info@aadcu.org

China Architecture & Building Press
www.china-abp.com.cn

In collaboration with:
Hodgetts+Fung
www.hplusf.com

d-Lab & International Architecture Research

School of Architecture, Central Academy of Fine Arts

Curator/Editor in Chief:
Bruce Q. Lan

Translation:
Keren He, Yufang Zhou

Book Design:
Hodgetts+Fung
Design studio/AADCU

ISBN: 7-112-07523-8

©本书所有内容均由原著作权人授权美国亚洲艺术与设计协作联盟编辑出版，并仅限于本丛书使用。任何个人和团体不得以任何形式翻录。

出版事务处：
亚洲艺术与设计协作联盟／美国
www.aadcu.org
info@aadcu.org

编辑与出版：
亚洲艺术与设计协作联盟／美国
bj-info@aadcu.org

中国建筑工业出版社／北京
www.china-abp.com.cn

协同编辑：
霍德盖茨＋冯
www.hplusf.com

国际建筑研究与设计中心／美国
中央美术学院建筑学院／北京

主编：
蓝青

协调人：
洛宾·罗，斯坦福大学

翻译：
何可人，周宇舫

书籍设计：
Hodgetts+Fung
Design studio/AADCU

SERIES OF CONTEMPORARY ARCHITECTS STUDIO REPORT IN THE UNITED STATES

Hodgetts
+Fung

目录

6	迷惑与目眩:霍德盖茨＋冯关于场所的即兴演奏	
	科特·W·福斯特	
12	身处其中的艺术	
18	道威尔图书馆，加利福尼亚大学	1991~1992
26	ZKT太阳能展览	1994~1995
	肌理	
34	穆林雕塑工作室，西方学院	1994~1997
38	查尔斯和雷·埃姆斯的世界展览	1994~2000
44	帕尔巴之城	1995
52	美国电影礼堂，埃及剧院	1995~1999
60	微软馆	1996
64	微软馆	1997
68	ZKT波状能展览	1997~1998
72	环球体验	1997~1999
	山野学校	
78	布鲁克林公园露天剧场	1999~2000
82	好莱坞碗形剧场	1999~2004
88	辛克莱馆，艺术中心设计学院	2000~2001
96	世界储备和贷款银行	2000~2002
100	萨尔玛图书馆	2000~2003
	细部	
110	虚拟住宅	2001~2007
114	海德公园图书馆	2001~2004
120	坦佩视觉与表演艺术中心	2002
126	加利福尼亚捐赠协会	2002
	生态世界	
132	国会图书馆／艾拉·格什温画廊，沃尔特·迪斯尼音乐厅	2002~2003
136	加维亚公园	2003
142	东京山野学校	2003~2007
148	建筑师小传	
150	主要参考书目	
152	精选项目	
153	精选奖项	
155	致谢	

Contents

6	Bedeviled and Bedazzled: Hodgetts+Fung's Improvisations on Locations Kurt W. Forster	
12	The Art of Being There	
18	Towell Library, University of California	1991–1992
26	ZKT Sun Power	1994–1995

Textures

34	Mullin Sculpture Studio, Occidental College	1994–1997
38	The World of Charles and Ray Eames	1994–2000
44	La Città Pulpa	1995
52	American Cinematheque at the Egyptian Theatre	1995–1999
60	Microsoft Pavilion	1996
64	Microsoft Pavilion	1997
68	ZKT Wave Power	1997–1998
72	Universal Experience	1997–1999

Yamano Gakuen

78	Brooklyn Park Amphitheatre, Minnesota Orchestral Association	1999–2000
82	The Hollywood Bowl	1999–2004
88	Sinclaire Pavilion, Art Center College of Design	2000–2001
96	World Savings and Loan	2000–2002
100	Sylmar Library	2000–2003

Details

110	Virtual House	2001–2007
114	Hyde Park Library	2001–2004
120	Tempe Visual and Performing Arts Center	2002
126	The California Endowment	2002

World of Ecology

132	The Library of Congress/Ira Gershwin Gallery at Walt Disney Concert Hall	2002–2003
136	Parque de La Gavia	2003
142	Yamano Gakuen	2003–2007

148	Biographies
150	Selected Bibliography
152	Selected Projects
153	Selected Awards
155	Acknowledgements

Bedeviled and Bedazzled
Hodgetts+Fung's Improvisations on Locations

Popular wisdom has God and the Devil contend one another over the details. What resides in the junctures is what holds the parts together. The joints, hinges, and edges of those parts are as precarious as the substance of the soul. If anything breaks (and hearts also do...), it is likely that the soul will fall apart too, since things with a soul break as easily as those that have none. But when they do go to pieces something of greater moment seems to crack.

What has been made instead of grown is almost always fitted together from parts. Joining makes an artifact. Even nanotechnology shuffles the particles of matter itself into new configurations. Buildings are such artifacts, locking together many parts and reaching such dimensions as the frame of our existence demands. Buildings can (and do) vary greatly: some are as small as doll houses, others as large, even gigantic, as aircraft hangars, factories, and bridges. Size matters, for the larger a building is, the more we need to shrink it in our imagination if we want to understand how it is made and how it works.

But size alone yields a poor measure, because modestly-scaled buildings can be so intricate as to defy one's imagination, while huge ones may be disappointingly simple. The large and the small coincide in the abstraction of a model. Ideally, models are made of few materials, none of them being those used in the actual building. Nothing appears in the model as it will in reality, though the rationale of its assembly and its intricacies are all brought within scope. True models—in contrast to small reproductions—incarnate the idea of a building rather than its body.

Buildings are never of a piece, even an igloo is put together from slabs of snow, and a pergola from timbers and bolts. Like the igloo, buildings tend to go the way of all things, sometimes decaying slowly, sometimes being demolished in a hurry. What they do not take away with them as they disappear is the *idea* of the igloo or the pergola. It is these ideas that turn a mere assembly of parts into architecture. If there is a difference, even a fundamental one, between buildings and architecture, then it lies in the fact that buildings rarely outlast their utility and their materials, while architecture lives on as an idea in memory.

As something that exists beyond its parts, architecture still cannot come into being without them. While its heart continues to belong to the daddy of its ideas, it still makes a play for the Caddy: Architects need not be whores—as has been claimed with cynical defiance—but they do have to know the business. And what a business it is, requiring more hands and more things, more tools, technologies, plans, and parts than almost anything else outside of the operating room and the space launch. For the uninitiated, a building site remains a confusing place, and for the professional, one of the most expensive, complicated, and frustrating—matched only by its own travesty, the cinematic set. What the two share in common is a product of such illusory nature that while one sinks huge amounts of money into pure imagery, the other sublimates major investments in enduring impact.

Imagine standing on a street corner waiting for the bus. As you idle away your time, you look around and notice a cement wall bordering the sidewalk, plants and shrubs above it filling the setback to a building's plate glass facade. Only shadows hover inside and the glass reflects the onlooker. As the retaining wall switches back from the corner, it folds down into a bench, formed of a thin slab with smooth edges. Only a few feet across, the built-in bench abuts another setback. Empty paper cups, a straw, and coffee stains on the sidewalk betray the passage of frequent bus riders who loll in the outdoor "nook" of the library. Readers have left even more obvious tracks. The entrance is at the far corner of the building, halfway between the parking lot and the sidewalk, where sliding doors give into a vestibule. A slanting wall of glass stands behind another, opening a scissor in space and locking it overhead with heavy cross beams.

As one enters off the sidewalk, rising a step or two, the cement gets smoother and cleaner, the light lower and the air cooler. Stepping inside feels almost as easy as folding the retaining wall into a bench at the street corner. Easy does it, and doing it is forgetting how unusual it is. Things are simple in this library and people come and go as if they were paying a call to a friend's house, shuffling back to the door or glancing at a poster, chatting all the way. The flow of readers eddies into nooks and spills into the children's corner. Nearby rooms await whatever use may be made of them. The foyer rises to the full height of the building, and its slanted walls seem to go straight through the roof.

This is only one of two branches of the public library system that **Hodgetts+Fung** have designed for the City of Los Angeles. It is located in a Latino community where books are scarce, but videos and newspapers, pamphlets and flyers surround patrons like a flock of gulls feeding on discards. The library turns into a trove, and the community embraces it as its own. As a building, it is numerous things: a quiet corner to wait for the bus, a tall, well-lit place to graze on the news, gravy for borrowers, and a cove to indulge the eyes. As architecture, the library shows how one can ply things, how they can be shifted and arranged before being locked into place. The smaller the parts, the more delicate they prove in the assembly of the whole, which is, of course, made of a great many different elements. Devilish details prey on the minds of the architects as they seek to save their ingenious ideas from being shredded by a mercilessly schematic translation into reality. Unassuming materials have long attracted **Hodgetts+Fung**, but unconventional ways of using them even more. If their buildings betray a workaday character, it is because they chose to fit them into that gray range, but add color and a special touch.

Although both of their libraries were tailored to their communities and convey a lot about life in two particular places in the far-flung geography of the city, they couldn't cut a stronger contrast. One is gracious in its gestures, the other more guarded—even somewhat forbidding. In the outlying Latino neighborhood of Sylmar, the library is open for business with a certain ease. In predominantly African-American Hyde Park, the library also stands its ground, if somewhat less easily. Here, many businesses line the street, most sporting hand-painted signs that give the passers-by a holler over the cottage roofs. The library doesn't stand back. On its facade, a ceramic wall gives free play to shapes assembled artfully from shards of glazed clay in the same hues as the painted signs. The tiles hand-made by local youngsters were shattered before their fragments found their place in the wall along the sidewalk. One may not find a window through which to peer inside, but the library is looking to the street with steadfastness, facing the public with a tableau from the hand of a local artist.

Hodgetts+Fung dare to tread on treacherous ground, be it socially precarious or technically difficult. They have recently restored to Angelinos one of their landmark sites, saving its familiar appearance while totally rethinking its performance. The Hollywood Bowl is an outdoor venue for music that can only work its magic with highly elaborate electronic amplification. The stage is framed in concentric arches whose acoustical properties have never helped their purpose, however much they have endeared themselves to concert-goers. With more than sleight of hand, **Hodgetts+Fung** have salvaged the familiar appearance of the music shed by inventing an intricate mechanism to pick up and disperse sound over the orchestra. Such suspended circular platforms have their unsuspected forerunner in the daring installation of a ring-shaped *cantoria* for singing youths which the Renaissance architect Brunelleschi hoisted under the dome of the Florentine church of San Felice in Piazza. Far from innocent in their guileless search for solutions, **Hodgetts+Fung** hit upon ingenious ways of handling even intractable problems.

Whether they hinge a door excentrically in a student lounge, with counterweights moving in calibrated motion, or mount an exhibition with brackets originally designed for altogether different purposes, they always restore a sense of amusement, in the original sense of the word suggesting serendipitous inspiration. When they need to conserve the image of an old movie theater while spiffing up its image and its technology, they zero in on the mechanics of their task and imagine ways of joining even incompatible parts. They obviously delight in chasing the Devil out of every impediment as they pursue a neat solution and add a glint to what they've taken in hand.

Hodgetts+Fung love to put things together, at home and in their office, in their heads and in their buildings. Atop a high ridge dividing the Los Angeles basin from the "Valley," they made themselves at home by putting discarded things to use: dismantled lettering from commercial signage is piled up to hold a glass table and a salvaged mechanical loom holds a collection of vintage vinyl LPs. The aluminum shelves of a highboy are held in place by industrial rubber bands. Only an inveterate tinkerer able to improvise a solution to any mechanical problem would stubbornly continue to drive a forty-year-old Italian car whose smoothly sculpted body has no use for awkward bumpers and no tolerance for potholes.

The art of assembly is the *techné* that answers to a spontaneous need for something practical, but originally sprang from the knack for putting things to (new) use. **Hodgetts+Fung** have been especially successful with temporary buildings and with places of spectacle. For an exhibition that reveled in the idea of assembly, Blueprints for Modern Living, Hodgetts+Fung found material after their heart's desire. Circumambulatory in its layout, the show sent visitors on a path through the post-war opportunities that turned whatever had been produced to new and often surprising use.

Adapting to a spirit that briefly spurred the post–war generation, our architects not only inhabit a genial glass house of the 1950s, but also cling to a spirit of exploration induced by a plethora of industrially-produced things. Impersonal objects, synthetic materials, and semi-finished products are the stuff of their imagination and fuel their ideas, but the heart also plays its part so that they don't turn out brittle or cold. Harking from the era of obsolescence, **Hodgetts+Fung**'s greatest strength lies in their capacity to rejuvenate things and improvise a future for them. In that old game with the Devil, they never cease to come up with new tricks and sly ways of keeping out of trouble.

Kurt W. Forster

迷惑与目眩：霍德盖茨+冯关于场所的即兴演奏

科特·W·福斯特

在公众的智慧中，上帝和魔鬼之争存在于细节之中。处于中间的东西起着连接各元素的作用。接缝、合页以及这些元素的边缘都是和灵魂本体一样不稳定的，如果有什么元素损坏了(本体中心也会损坏)，就很有可能灵魂本体也会被破坏。虽然有灵魂的和没有灵魂的一样脆弱，但是当它们真的成为碎片的时候，一些更重要的东西就可能不存在了。

被制造而不是生长出来的东西几乎都是由元素组成的，经过连接后而形成人工制品。即使是微分子技术也是把物质的粒子重组，形成新的组合。房屋是人工制品，它把许多部分锁定起来，形成一定的人们生存所需求的尺度。房屋可以(也的确)有很大的不同：有些小如娃娃屋，有些则很大，甚至像停机库、工厂和桥梁一样巨大。尺寸是很重要的，因为一个房屋越大，我们就越需要把它缩小到我们的想像范围内，只有这样我们才能理解它是如何建造和运行的。

但是仅用尺寸作为衡量标准是很糟糕的，因为一个中等尺寸的房屋可以复杂到超出人们的想像力，而一个巨大的房屋则可能简单到令人失望。大和小的概念都能在模型中提炼出来。理想地说，模型是由少量的材料制成，这些材料都不是用于真实房屋的材料，在模型中展现的一切也不会在现实中成立，但是房屋组成的逻辑关系和它的复杂性却能在此表现出来。和这种小比例复制品相对比的真实的模型，则是把房屋的理念具体物质化而并非只是复制它的形体。

房屋从来不只是一个部分，即使爱斯基摩人的冰屋也是由雪块、木料和铆钉制成的棚架所构成。像冰屋一样，房屋的所有部分都会逐渐消失，有些部分缓慢地腐化，有些则很快地被拆除。但没有随之消失的则是冰屋或是棚架的概念。就是这些概念使它们成为建筑而不仅仅是一些物质的组合体。如果房屋和建筑之间有差别，哪怕只是最基本的，那就是房屋的用途和材料很少持续耐久，而建筑则是能存留于记忆中的一个概念。

虽然建筑的存在超越了它的组成部分，但是没有它们，建筑也不可能实现。即使在内心深处忠于理念，它们也会玩现实的游戏：尽管建筑师不需要变成娼妓，正如一些愤世嫉俗的人蔑视的那样，但是他们也应当懂得做生意。他们的这种生意需要更多的人力，更多的工具、技术、计划和部件，除了手术室和航天发射以外，他们需要的东西比任何其他行业都多。对于外行来说，建筑工地一直是一个令人迷惑的所在，对专业人士来说，是一个最昂贵、最复杂和最头疼的地方，只有它的拙劣仿制品——电影摄制棚，可以与之相媲美。两个场所的相同之处在于它们都是视觉想像的产物，不同之处在于一个是把大量的金钱投入到纯粹的想像中，而另一个则是使主要的投资升华形成一种持久性的影响。

设想你站在一个街角等汽车，当你为打发时间而四处张望的时候，你注意到人行道边的一堵水泥墙，上面的树和灌木一直填充到退在后面的建筑的玻璃立面前。玻璃里面只有阴影，反射着观察者的形象。当挡土墙从街角转过来的时候，向下折叠变成一个坐凳，由一块薄薄的边角光滑的板制成。坐凳只有几尺宽，衔接着另一边退后的红线。空纸杯、一个吸管和人行道上的咖啡印记暴露了经常等汽车的人们徘徊于此的行迹。来看书的人则留下了更明显的痕迹。入口在房子的远处转角，在停车场和人行道之间一半的距离内，从滑轨门进入一个门厅。

一片倾斜的玻璃墙站在另一片玻璃墙的后面,在空间形成一个剪刀状,在屋顶处用沉重的交叉梁锁定。

当你从人行道下来,走上一两步台阶,水泥墙变得更光滑干净,灯光更低,空气也更凉爽。进入室内就好像那堵挡土墙在街角变成坐凳那样轻松;看上去轻松,做起来就是要把它有多不寻常忘掉。图书馆里的一切都很简单,人们来来去去就好像给朋友家打电话一样,在门口驻足,看看海报,不停地聊天。读者人流在角落里盘旋或是涌入儿童阅览部分。旁边的房间可以用作各种用途。门厅是通高的,它的倾斜墙看上去像是直接穿过屋顶。

这是霍德盖茨+冯事务所为洛杉矶市设计的两个公立图书馆之一。它坐落在拉丁族聚居区,书籍不多,但是影像带、报纸、手册和传单包围着一群如觅食海鸥般的读者们。图书馆成为了一件珍宝,而社区接纳它为其一部分。作为一幢房子,它是很多东西:一个等公共汽车的安静角落,一个高大明亮的看新闻的地方,借阅者的意外财富,或是一个饱眼福的角落。作为建筑,这个图书馆体现了建筑师如何使用物质,怎样在把东西锁定在空间之前转换和安排它们。元素越小,就被越精细地表现在总体中,这个总体当然是由许许多多不同的元素所组成。当建筑师们在把想法付诸于现实的时候,总是设法从无情的现实中拯救他们的天才灵感,同时极度的细节就会产生于他们的头脑中。霍德盖茨与冯长期以来一直对不常用的材料感兴趣,并且用不寻常的手法来运用。如果他们的作品暴露出乏味的特点,那也是他们意图融合那种灰色的范围,但不忘添加上一些颜色和手法。

尽管他们设计的两个图书馆都是为他们所在的社区度身定做的,它们在距离很远的城市的两端传递了很多的信息,在这点上,这两个图书馆所形成的对比是极其强烈的。一个落落大方,一个则保护性很强,甚至有点森严。拉丁族社区的萨尔玛图书馆运营起来有一定的轻松感,而在以非裔为主导的海德公园社区,图书馆也是同样稳稳地站立着,却不是那么轻松。在这里大多数商业建筑临街而设,许多沿街商业建筑在屋顶上都有夸张的手工绘制的招牌。此图书馆也没有退街道红线,他的立面的一面墙由形状自由的釉面砖的碎片组成,砖的颜色和周围的手工招牌相一致。瓷砖是由当地年轻人手工制成,打碎后在图书馆临街的立面上拼贴起来。人们也许找不到一扇窗户可以从街边往里面看,只有图书馆默默地注视着街道,对着公众的是一幅由当地一个艺术家绘制的戏剧化场景。

霍德盖茨+冯事务所敢于进军不安全的领域,既有社会意义上的不稳定性,又有技术上的难度。他们最近改造了美国人的一个标志性场所,保留了它的外表而同时重新思考了它的用法。好莱坞之碗是一个室外音乐剧场,只有在强力的电声放大下才能现它的魅力。舞台包容在同心的圆弧下,不管听音乐会的人多么倾心于这个形式,对声学来说起不到任何作用。经过很小的改动,霍德盖茨与冯保留了这个音乐舞台的外表,并且发明了一个复杂的机械装置,可以把乐团的声音收集起来并扩散到听众席去。这种悬挂的圆环有据可查,文艺复兴时期的建筑师伯鲁乃列斯基(Brunelleschi)就在他的佛罗伦萨穹顶下,为唱诗班大胆地装置了这样一个华盖。霍德盖茨与冯不只是在寻找诚实的答案,他们更多地是在为难以驾驭的问题设想天才的解决方法。

在学生中心他们把门偏心安装，重锤按刻度移动，或是用与初始目的完全不同的支架来安装展板，他们总是用原创的偶发灵感来表达一种幽默感。当他们需要注入新的设计和技术，同时又要保护一个老电影院的形象时，他们把技术设备用在最不显眼的地方，并且设想出把不兼容的部分结合在一起的方法。他们很明显地喜欢寻求一个干净利索的解决问题的方法，驱散遭遇难题所带来的不快，并且同时增添一些闪光点。

霍德盖茨与冯喜爱把各种物体结合起来，无论是在家里还是在办公室里，在他们的头脑里或是在他们的房子中，在分界洛杉矶盆地和"山谷"的山脊上，他们游刃有余地把一些废弃的物品重新利用：办公用剩的信纸被他们堆积起来支撑一个玻璃桌子，一个收集来的机械织布机被用来摆放一些经典的乙烯LPs。高橱柜的铝质隔板被工业用橡皮筋固定着。这好似只有很熟练的、能随心所欲地解决任何机械问题的修补匠，才会固执地开一辆使用了四十年的意大利车，车身光滑而具有雕塑感的老车无需保险杠，也经不住坑凹路面的颠簸。

装置艺术是一种对一些实用性的东西产生自发性的对策的技巧，但是最早是来源于改造旧东西，巧妙地赋予其新的用途。霍德盖茨与冯很擅长于临时性的、奇特的建筑。在一个有关装置艺术的展览——"现代生活的蓝图"上，他们找到了寻求已久的素材。展览成环形布置，参观者沿着战后的发展路线，可以有机会把任何东西改造成新奇的功能。

战后一代的建筑师受到的激励，不仅仅是1940年设计的玻璃房子，还有过量的工业产品所带来的影响。非人性化的物品、合成材料和半成品充满了他们的想像，融入了他们的思想。然而他们内心仍是活跃的，因而才不至于变得冷酷和暴躁。霍德盖茨与冯从过去的时期汲取灵感，他们最大的力量在于他们对旧东西的更新，并赋予其新奇的未来。在避开麻烦的游戏里，他们从不停止新的花招和淘气的躲避。

The Art of Being There

With the digital order in the ascendant, massive transformations have occurred in nearly every aspect of our lives. These have triggered a renewed interest in the obscure and the arcane, low-tech and "slow" enterprises, while at the same time accelerating the trend toward devices and environments defined by digital technology. However, like it or not, the human body is an analogue device. We can neither see, hear, nor touch digital information without a prosthesis—a screen that translates a stream of digital Xs and Os to produce an image, or a speaker cone driven by an analogue signal derived from these Xs and Os. Our world, it turns out, is unremittingly analogue. We see that the glass is half empty and intuit how long it will take to fill it. We note the size of a container, assess its structure, and anticipate that it is either heavy or light. We do this with rooms and spaces, with staircases, chairs, drawers, and windows. Our first-person encounter with the world around us, unmediated by interpretive devices, defines our experience, and will, we suspect, well into the foreseeable future.

We embrace the digital revolution. Its impact on technology, design, and lifestyle cannot be overstated, and we are mesmerized by the degree to which it promises to liberate architecture from constraints that originated with the repetitive order of the industrial revolution. Promising speed which was unthinkable in the proto-digital world of Morse Code and accuracy beyond that attainable with the protractor and yardstick, digital tools can accomplish in seconds what Eiffel's draftsmen labored endlessly to produce. Yet, with all that power, digital tools and systems do not enable us to think more profoundly about the merits of our work. It is important, above all, to distinguish the physical reality of a building, an installation, or an object from the means of its conception and realization. For us, the highest value rests in the way disparate materials are assembled to form a harmonious whole, or the manner in which an entirely new material technology finds its place in the hierarchy of spaces, geometries, and textures that frame experience. Our work has its roots in experience, rather than in any sort of formal principles or processes—digital or otherwise. Governed by the eye of the viewer and the dynamics of motion, our aim is to design spaces that impel physical interaction, injecting users into environments that re-animate the senses.

We are not interested in replicating the two-dimensional plane of the cinema or the plasma screen or the billboard. Nor are we satisfied with a picture-perfect environment whose sleek surfaces conceal no secrets. What interests us is the unique experience architecture shares with no other art, *the art of being there*, of being completely enveloped in spaces that are singular, appropriate, and purposefully orchestrated.

Observing cities and buildings, we find it impossible to separate foreground from background, building from context, or architecture from artifact. It was ever thus. Whether the street is filled with horses or automobiles, lined with telegraph cables or specimen trees, the eyes of people with their feet on the ground take in all things as a single, complex retinal image. And while the digital processor is capable of defining edges, assigning dimensions, and plotting locations, it is the analogue mind that assigns value and meaning to our surroundings, seeks out the character of the environment, and secures our place within it. Analogue experience is all about being there.

Architects, like other artists, are either the masters or the victims of their milieu. Led by enthusiasm for the new or nostalgia for the past, the result of their efforts in any one period can be seen as a field of supple shoots that waft to and fro in the winds of cultural and technological change. This is not a bad thing. The gestation period of most buildings often results in the blurring of a decade or so of evolution. But given the rate of change driving today's cultural and intellectual engagements, and the superabundance of new materials that run the gamut from high-tech to recycled, it is a virtual certainty that any given building will no longer be framed by the same neighboring structures in a matter of ever fewer years. Mirrored glass that comfortably abuts pre-cast concrete will certainly have a new, perhaps less compatible neighbor in the proximate future, adding to the hurly-burly alchemy of the future urban environment.

The self-contained narrative that was once the province of heroic, freestanding architecture survives today for the most part in the museum, theatre, or concert hall, whose presence is intentionally monolithic and isolated in its refusal to merge with, or even recognize the existence of its urban surroundings. By contrast, in rejecting such an approach, our work represents a search for composite form and substance with which to embed the new in an existing urban framework. Distant views, the pulse of traffic, the immediacy of an historic artifact, and the delight of the user's sensibilities form a mosaic of references and inspirations that give purpose to our design. Like a system upgrade, these principles drive us to perform a nearly seamless intervention rather than proclaim the birth of a new order.

On the one hand, the often radical juxtapositions in our projects—extruded polycarbonate meshing with galvanized metal studs, or industrial steel grating bordering intricate faux-Egyptian motifs—reflect our philosophy of inclusion, and on the other hand, assert our conviction that the provenance of contemporary materials is no longer the main criterion for their use. In our practice as well as in our teaching, we assert the rise of a meritocracy in architectural design. Materials are no longer desirable primarily for their luxury status or exotic origin, nor is the hand of the artisan valued above the nozzle of a digital cutting machine. Mass-produced, off-the-shelf, and interchangeable components must pass through the same sieve as the refined and the rare before assuming a place in our buildings.

Given this level playing field, materials are chosen for their meaning, their aura, and their sympathetic relation with the mission of the project, but this does not mean that we seek refuge in the past. The past is over. The lifestyles, politics, and technology that created past forms no longer apply. However, the destiny of architecture and of our future cities, we believe, will not—indeed cannot—be fulfilled by the seamless volumes and voids of the digital domain. Those images, compelling and deluxe beyond imagining, have found a secure niche in product design, wherein the superficial appeal of a shape is encoded with both status and value, but they remain "objectified." We love to look at them, rather than inhabit them, and it remains to be seen whether our increasingly diverse spatial appetites will be whetted by those voluptuous folds. Architecture of our time must reframe the dialogue between perception and intellect, take a step beyond the mere appropriation of digitally-inspired forms, and assess the manifold forces that will mold the architecture of the future.

Everything from manufacturing and industry to logistics, retail shopping to office work, education to recreation is being retooled as we write, yet many architects focus on digital dreams to the exclusion of tangible environments that can have a substantive effect on the way we live and work, on human interaction and engagement. It is a paradox that digital architecture, so effortlessly glowing on the display panel, all too often assumes a self-defining, and we believe, self-limiting role. Our expectations are foreshortened by the seeming congruity of the surfaces, the reflections, the lighting, and the animated/stilted postures of the digital inhabitants. Rather than plumbing the realities of myriad potential scenarios, we "buy" the image. Extracting plan, section, and elevation, we proceed to build a simulacrum of the digital image from permanent, intractable materials, eschewing practical constraints and challenging conventions.

The packed shelves of the merchandise mart, the endless commute, and the bleak stacks of office cubicles are about to be consigned to the dust bin of antiquity, as the compact appliances of future tasks mold themselves to our bodies, interface with our tasks, and guide our hands and machines to a more and more eloquent dance with substance. The surfaces once needed for writing, drawing, perusal of charts and maps, and manipulating data have outlived their usefulness. Likewise, from a certain perspective, the problem of high-density information storage and retrieval has been liberated from the accompanying maze of aisles lined with the bookshelves, file cabinets, desks, tables, and lighting that we came to take for granted. What is left in both dwellings and workspaces are informal social spaces in which the configuration of walls and furniture is no longer defined by functionality, but rather derives from the congeniality of the space itself. Ambience, lighting, and the sensual, tactile qualities of social space need no longer be mitigated by the stubborn rectilinearity of a storage unit. Freed from the burden of gravity, floors and walls are able to assume new roles, flexing to offer better places to live, work, play, or simply relax.

In subtle ways, these changes are insinuating themselves into our lives as well as our work—as our work and our lives become utterly indistinguishable. As technology advances, spaces once cluttered with devices are purified; spaces for hygiene and sustenance, once deemed purely "functional," become more seductive, while formerly "dedicated" spaces are assigned provisional functions determined only by the nature of the portable accessories one brings to them. Some may argue that the average American family has become impossibly passive, immersed in food and TV. But we see it another way. We see American culture as dominated by an unfettered desire to escape conformity. The rhythms of hip hop, the jangle of bling bling, and the mechanics of the Internet jab at the seamlessness of the new aesthetics. It is clear that priorities are molding themselves to a new order in which the body and the environment are engaged in a primal *pax de deux*. Stripped down enclosures, shorn of specialized functions, are reassigned to a vague ceremonial or social mulch, perfect for *poseurs* and their *louche* companions.

Here we see an opportunity for a new architecture, as an accompaniment rather than a container, in which each space is a library of resources, differentiated by ambience rather than function. After all, if any space might be appropriated for any function (and present technology suggests that this will be the rule, rather than the exception), then it falls to the elements of a particular space to present an opportunity for transient use. The fit of enclosure to purpose might then parallel the meshing of the emotional resonance of the space to the tenor of the transaction.

We have built our design philosophy on the holistic view that every aspect of our surroundings, from tableware to thoroughfares, is a participant in the human experience, and that the architect, as master of the environment, has the right and the obligation to employ this arsenal of experience in the creation of a compelling environment. Imagine an architecture in which the mission is to achieve difference rather than uniformity, in which the hand of the architect confers a vigorous dissonance rather than a polite accord. We no longer ask our artists or musicians to play within a confining set of rules. Why, then, our architects? With the freedom to download and remix, to morph from streetwise to all business, to flaunt tattoos and piercings, SUV's and skateboards, our culture has staked out an age in which identity is forged by the individual who assumes it. Why not employ architecture as a catalyst, rather than a constraint? Why not embed the behavioral cues, the linguistics of posture, attitude, and action in the fabric of our structures, rather than pursuing an agenda of formal prescription? The diversity of our work attempts to confront this challenge.

Craig Hodgetts + Hsin-Ming Fung

身处其中的艺术
克雷格·霍德盖茨＋冯心明

随着数字化秩序的增强，在我们日常生活的几乎每一个方面，大规模的转化正在进行着。这种潮流激发了一种新的对那些模糊的、神秘的、低技术的、以及"缓慢"的产业的兴趣，并且在同时也加速了那种倾向于数字化技术设备及环境的潮流。然而，不管你喜欢与否，人体是一个类比性的装置，我们不能看、听或是触摸数字化的信息，而是要通过一个媒介、一个荧光屏，把一组数字的×和○转化成图像，或是由×和○所产生一个模拟信号来带动扬声器。我们的世界竟变成一个模拟的世界。我们看到一个半满的瓶子，然后分析出用多长时间能灌满它，我们注意到一个容器，估计它的结构，然后分析出它分量的轻重。于是我们用房间、空间、楼梯、椅子、抽屉和窗户来进行同样的分析。第一个接触我们周围的这个世界的人，在还没有被其他诠释性的装置影响之前，定义了我们的体验，并且我们猜测这种体验将会被完整地带入到可预示的未来。

我们拥护数字化革命。它对技术、设计和生活方式的影响不容忽视。在某种程度上我们曾承诺过摆脱由于工业革命的重复秩序所带来的对建筑的束缚。承诺在速度上，那种莫尔斯密码和电报时代所无法想像的速度；承诺在精确度上，那种超越了场地标志物和划分线的精确度；承诺的还有数字化的工具，埃菲尔的绘图员花费无数时间的成果在几秒钟内就可以达到。尽管数字化工具有如此强大的力量，但是它并不能使我们更深入地了解我们作品的优点，而且最重要的、高于一切的是分辨出一个建筑的实体特性，一个装置，或是从概念和认知的手段上得出的一个物体。对于我们来说，最高的价值体现于把不同的材料组合成一个和谐整体的方法，或者是把一种崭新的材料技术在体验空间、几何和材质的秩序中合理地运用。我们的作品植根于体验，而不是什么其他正统的数字化或者非数字化的原理和过程。由普通人的观察和活动所驱使，我们的目标是设计能激励人们身体力行的交流空间，刺激身处其中的使用者的感官。

我们设想的不是模仿二维的电影放映屏幕、全视角荧光屏或是广告牌之类的东西。我们也不满足于仅仅描绘一个上镜的环境画面，肤浅的表面其实一览无余。我们感兴趣的是创造一种特殊的体验，是建筑不同于其他艺术所特有的一种身处其中的艺术。这种感觉是完完全全被包裹在一个独特的、合理的、有目的地组织起来的空间中。

纵观城市和建筑，我们认为无法把前景和背景分开，把房屋和文脉分开，或是把建筑和工艺分开。不管街上走的是马车还是汽车，是架着电线还是行道树，从来没有改变过的事实是，站立在地上的人把用眼睛所包容的一切看作是一个独立的、复杂的画面。当数字化工具可以分割边界，标注尺寸和打印位置的时候，是类比的想像定义了我们周围环境的价值和意义，探询了环境的特性，然后把我们的位置安排其中。类比的体验本身就是身处其中的艺术。

建筑师——就像其他艺术家一样，或是他们自我世界的主人，或是受害者；或是被新事物的热情所驱使，或是被怀旧的情绪所影响。他们在任何一个阶段所努力的结果都可以看作是随着文化和技术的变化之风而摇摆。这不是一件坏事。大多数建筑的设计孕育时期都导致了一个或几个世纪的进化。但是促使当今世界文化和意识概念变化的速度是如此之快，加之全范围的、从高科技材料到回收材料的极度丰富的新材料选择；任何一个房屋周围邻里结构的改变，比起以前时间要短得多，这已经是不争的事实。镜面玻璃的房子可以很舒服地与预制混凝土房子相毗邻，它们将来的邻里环境肯定是一个较新而不太融合的结构，在未来都市环境喧嚣的融炉中冶炼着。

以自我为中心的表述在以前曾是英雄式的独立建筑的领地，而今天则大多用于博物馆、剧院或音乐厅。这些建筑的存在是内在地以自我为中心，孤立地拒绝甚至不承认周围环境的存在。与之相反，我们的作品拒绝这种意向，而是表达了一种探求能融合到现有都市环境中的综合的形式和本质。视角研究、交通规律与历史文物古迹的关系，以及使用者感官的愉悦，都构成了我们设计目的的参考和灵感。就像对一个系统的更新，这些原则促使我们形成一个几乎不露痕迹的表现，而不是宣称又发明了一种新秩序。

在一个方面，我们作品中常常把一些元素并列使用，比如说模制的聚酯碳酸网和镀锌金属龙骨，或是厂家生产的钢制边框镶着复杂的仿埃及图案，这些都表达了我们的包容一切的理念。在另一个方面，则确立了我们的信念，那就是当今时代，材料的出处不再构成使用它的一种标准。我们在实践和教学中都坚信建筑设计的贤能政治时代的开始，材料不再会因为它们的奢侈品质或是新奇的来源而提升价格，或是数字机器切割的产品比手工制品低廉。大批量生产、异地生产和可互换的产品和那些精致稀有的材料一样都经过筛选过滤后才能用在我们的设计中。

在这个基础上，材料被选用是由于它们的意义、气氛以及它们和设计的和谐的关系。这并不表明我们寄存在过去的氛围中，过去的已经过去了，过去的生活方式、政治和技术已经不适用了。但是我们相信建筑的命运和我们未来的城市不会，也不可能完全由数字化控制的毫无缝隙的体积和空间所构成。这些数字的画面如此刺激和豪华，以至于超越了画面本身，在产品设计中找到了很安全的一席之地，在那里，表面的形状代表了产品本身的地位和价值。然而它们仍是"物体化"的，我们也许愿意观赏它们而并不愿意住在里面。无论我们对空间多样化的胃口是否能被那些刺激欲望的形状激发起来，这些物体仍处于被观赏的地位。我们这个时代的建筑必须重新建立起视觉和思维的对话，在纯粹以数字化形成的形体之上更进一步，再加上对多种影响因素的分析，这样才能塑造出未来的建筑。

我们重新书写所有的一切，从生产制造到后勤管理，从零售商业到办公作业，从教育到休闲。然而许多建筑师将注意力集中在数字化的梦想中，认为它能解决一切我们周围可触摸的环境问题，而这些环境对我们的生活和工作、人类的交流和互动都有着可持续发展的影响。这是一个矛盾，因为数字建筑那么不知疲倦地在荧光屏上成长着，我们认为大多数情况下暗示了一种自我定义，更不如说是自我限制的角色。我们的期望被那些看上去连续的表面、反射、灯光和生硬的动画人给简化了。在理应照顾到我们那些无穷无尽的可能性情景的时候，我们却简单地相信了这些数字图像，然后从中生成了平面、立面和剖面。我们从数字图像开始建造一个幻象，采用那些最难对付的、永久性的材料，而避开实际的功能限制和有挑战性的条例。

排列得密密麻麻的超市货架、冗长的上下班交通，和那些空空荡荡的堆积在一起的办公室桌位，都渐渐要被扔到古董店里去了。同时未来任务被压缩的数字化设施被我们度身定制，干涉着我们的工作，带引着我们的手和机器越来越熟练地操作着。曾经这些都是需要我们去写、去画、去查地图和表格，去分析数据使之为我所用。就好像从某一个角度来讲，高密度的信息储存和提取使我们从无穷无尽的书架、文件柜、书桌、桌子和灯光之中解脱出来,而这些曾是我们认为很重要的东西。我们居住和工作所剩下的只是一些不正式的社交空间，被已经不具有功能性的、只是由空间本身生成的墙和家具所围合。气氛、灯光和那些感性的可触摸的社交空间不再需要被一字排开的储藏单元所减弱。摆脱了重力的束缚，地面和墙都可以有新的功能，自由又灵活，能变成更好的起居、工作、娱乐或只是简单地放松的场所。

以某种微妙的方式，这些变化正在巧妙地进入我们的生活和工作，其实我们的工作和生活也正在变得界限不清。由于技术的进步，曾经需要配备很多设施的空间被净化了，卫生和维修保养的空间曾被认为是纯"功能性"的，现在则变得更加有诱惑力；而曾经是"指定功能"的空间则只用作暂时性的功用，在里面放置的可携带的东西则决定了它们的功能。有些人可能争辩说平均数量的美国家庭正变得不可救药地被动，沉溺于食物和电视之中。而我们却不这么看，我们认为美国文化正被一种不受束缚的、试图摆脱标准行为的欲望所控制着。嘻哈的节奏、Bling-bling的丛林，以及互联网的机械部件都变得天衣无缝，新奇美观。可以很清楚地看出，生活的要素正在适应一种新的秩序，在这种秩序里，身体和环境交融成一种首要的两重性。外表被剥开，特定功能被撕掉，空间被赋予一种模糊的、纪念性的或是社交性的属性，正好为那些矫揉造作的人群服务。

于是我们看到了创造一个新建筑的契机，如同一个附属物而非一个容器，每一个空间都是一个资料图书馆，以气氛而非功能来分别。毕竟，如果任何一个空间都可能用作任何一种功能（现代技术建议的是这条规则，而不是相反），那么就推导出一个特定的空间的特征就是提供一种短暂应用的机会。把围合空间的外表与目的相统一，就如同调和富有感情的空间共鸣和其中发生行为的高音一样。

我们以一种全方位的视角来建立设计理念，我们周围的所有元素，从桌上用品到城市干道，都是人们体验的一部分；所以建筑师——作为环境的主人——有权利和义务来利用这些贮藏的体验，创造一个有激情的环境。设想一个建筑，它的目的是创造不同而不是统一；建筑师的工作是探讨一种富有活力的不和谐，而不是有礼貌地和谐。我们不再要求艺术家和音乐家这样做，那为什么如此来要求建筑师呢？随着自由地下载和重组，从街头到所有的商业，夸耀纹身和穿孔，SUV和滑板，我们的文化正面临着一个时期，个体识别性是由个体的假设所塑造成的。为什么雇一个制造混乱的建筑师，而不是一个有约束力的呢？为什么不把行为尺度、身体语言、态度和行动编织到我们建筑的材料里去，而是循规蹈矩地照方抓药？我们作品的多元化特性正是试图面对这样一种挑战。

A Hodgetts+Fung building is never pristine, never inevitable. Their buildings move, change, or simply appear as if they aren't quite ready to settle into stasis. More than almost any other American architectural practice, the work of Hodgetts+Fung seems to embody and even perpetuate the notion of process. In their hands, technology is almost always deployed to engage the user rather than merely to solve a technical problem. Recognizing that left to its own devices, the technosphere is a lonely place, Hodgetts+Fung embrace technology—but not as an end unto itself. On the contrary, these inveterate humanists at play in the technosphere understand that technology has the power to enable human interaction. The same is true of every one of the projects presented in this book. At a time when technology so easily breeds isolation and anomie, Hodgetts+Fung offer us an architecture of hope. - **Reed Kroloff**

TOWELL LIBRARY
UNIVERSITY OF CALIFORNIA
LOS ANGELES, CALIFORNIA 1991–1992

道威尔图书馆
加利福尼亚大学
洛杉矶，加利福尼亚 1991~1992

The Towell's mission was to temporarily serve the comprehensive functions of UCLA's Powell Library, one of four original campus buildings built in 1929, while a five-year seismic retrofitting operation was underway. Severe time and budget constraints, together with the university's stipulation of no permanent disturbance of the designated "plaza" site at the foot of Janss Steps, a formal staircase ascending along the axis of the historic quadrangle, dictated the use of a proprietary tension structure. **Hodgetts+Fung**'s design for the Towell was conceived as a grouping of tented "land-forms" arranged to receive the Westwood extension of the original campus axis and redirect it toward the busy student center to the south.

霍德盖茨+冯的建筑从未拘于传统,从来都是不可预料的。他们的建筑在运动、变化或者就像是在顺其自然的过程中显现,如果它们还没有进入到一种停滞的状态中。霍德盖茨+冯的作品与大多数的其他美国建筑师的不同之处在于——他们的作品看起来被赋予了在过程中求得永恒的观念。在他们的手中,科技总是在与使用者的碰撞中得以展开,并超越了对技术问题的简单解决。其中引人注目的是技术范畴作为一个独自的领域,而不是基于其自身的解决方案,霍德盖茨+冯接纳了科技,但不是对科技本身的终极追求。相反的,对技术的理解是这些根深蒂固的人文主义者的游戏——科技拥有促进人类交流的力量。这就是呈现于本书中的每一个项目的真相。当科技很容易引发隔阂和冷漠时,霍德盖茨+冯给了我们一个建筑学上的希望。雷蒂·克劳夫

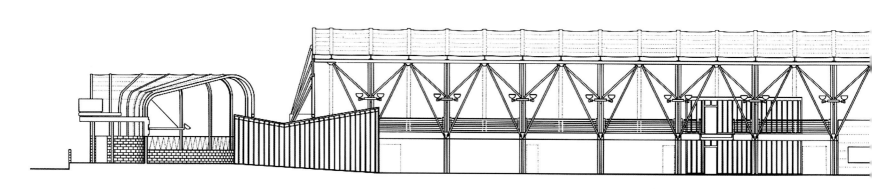

Flanked by the Men's Gym and the Dance Building, the Towell's curvaceous profile and audacious primary colors starkly contrasted with the historic faux-Romanesque red-brick buildings at the heart of the campus, with their porticoes, cortiles, and loggias. Seemingly casual and boldly asymmetrical, the Towell was nevertheless tightly woven into the architectural fabric of its surroundings with playful homage to the school colors blue and yellow, and subtler references to the curving balustrades and banded masonry of the existing buildings. Despite its ephemeral nature, this jubilant building rapidly became an icon on the campus, with its inviting parti and luminous nighttime presence.

Hodgetts+Fung employed a re-usable aluminum and fabric roofing system for rapid assembly, linking the major enclosures with an ensemble of both disposable and re-usable masonry, wood, and plastic substructures. An integrally-colored concrete block evocative of the brick-and-limestone work typical of the historic campus buildings establishes the periphery of the land-form, suggesting the foundations of a building long-since gone, and reinforcing by contrast the temporary nature of the superstructure. Lightweight, with a skeleton akin to that of an aircraft, the Towell's regularly-spaced aluminum ribs were covered with a stretched-fiberglass membrane that was in turn anchored with laced cables to surrounding masonry walls. This was a building that yielded to the elements, moving as much as a meter under gale force winds, and expanding and contracting with changes in temperature. A clerestory of overlapping plastic shields accommodated these changes, allowing the slender roof structure to move while the more robust supporting structure, housing books and serving the research needs of students, remained in place.

On the inside, a cavernous space was criss-crossed by struts and illuminating fixtures. Arranged to emphasize the rhythmic spacing of ribs and translucent panels, these doubled columns leaned outward to receive the roof structure then angled vertically to support a heavily laden mezzanine. Muted light tinged with the blue paint of exposed studs was admitted through pressed fiberglass panels at each end of the building, flooding the interior with daylight, and creating a softly glowing beacon at night. To the east and west of the main space lay two smaller pavilions, both circular in plan, linked by glowing passageways called "hammocks" for their relaxing nature. The "core" of each pavilion was composed of a cluster of ribs which served to funnel torrents of rainfall to below-grade conduits , as well as supporting the "velum" of the roof.

当洛杉矶加利福尼亚大学的鲍威尔图书馆(建于1929年,是四个校园最早的建筑之一)要进行五年的抗震修复时,道威尔图书馆的建造就是为了临时性地代替它的综合性的作用。紧张的时间和紧缩的开支,加上学校规定的不允许长期影响杰斯台阶下面的指定广场(杰斯台阶是一组沿着早期的四方庭院轴线的纪念性大台阶),这一切限制最后决定了新图书馆用一种专门的悬索结构的可能性。霍德盖茨+冯对道威尔图书馆的设想是一组拉索的"地形",在布置上承接Westwood延伸的原校园轴线,然后改变其方向,引导到南边繁忙的学生中心去。

道威尔图书馆夹在男生体育馆和舞蹈馆的中间,它弯曲的轮廓和大胆的原始色彩和校园中心区的古典建筑形成完全的对比,后者是仿罗马风的,带着门头,内院和柱廊的红砖建筑。新图书馆看上去似乎是随意的,大胆而非对称的,但是它却紧密地结合在周围建筑的环境中,比如说它对蓝色和黄色的校色的尊重,微妙地参照了现存建筑的弧形栏杆和条纹形的外墙面石材。除去它的临时性的本质,这个欢快的建筑因为它吸引人的功能布置和它夜间的照明,很快地成为了校园的标志建筑。

霍德盖茨+冯采取一种可再利用的铝和织物的可快速组装的屋顶系统,它和主体的连接也是采用一些可再利用的石材、木头和塑料来做次结构材料。一种本身有颜色的混凝土块,使人联想到老校园建筑典型的砖和石灰石,构成了地形的边缘,暗示着建筑的基础早已消失,强化了上层结构临时性的感觉。图书馆外立面规则排列的铝质骨架很轻巧,类似飞行器的材料,罩上一层拉伸的纤维玻璃。这种纤维玻璃同时也用钢丝固定在周围的混凝土墙上。这是一个随自然驱使的房子,它在强风力推动下可移动一米的距离,随着温度的变化伸展或收缩。一条纵向天窗适应了这些调整,在支撑结构不动的时候,这条天窗使得轻巧的屋顶可以移动。

在内部,一个洞穴式的空间,穿越着不同的支撑构件和照明轨道。为了强调骨架结构和不透明面板的间隔,这些双柱向外倾斜,承接屋顶滑行的铁袄龙板,然后又垂直向上承接夹层楼板。静谧的光线微微染上涂成蓝色的龙骨,从建筑两端的纤维玻璃板进入,使得室内充满阳光,夜间照明则使建筑成为一个有柔和光亮的信号灯。主要空间的东西两端各有一个圆形的小亭子,由一条闪光的走道相连接,因为它轻松的特性而被称为"吊床"。每个亭子的中心是由一组龙骨构成,几乎像哥特结构一样,用作暴雨时引导雨水到地下管道的漏斗,同时也用于支撑半透明的屋顶板。

ZKT SUN POWER
THE ELEKTRIZITÄTSWERK MINDEN-RAVENSBERG (EMR)
HEADQUARTERS ENERGIE FORUM-INNOVATION
BAD OENHAUSEN, GERMANY 1994–1995

ZKT太阳能展览
EMR总部，新能量论坛
巴特恩豪森，德国 1994~1995

The study of solar power is as old as mankind. Sorcerers, priests, prophets, and magicians were the first to expand human understanding of the sun's nature. Then came the pre-Socratic philosophers, and later, astronomers and navigators, and finally, the physicists. For thousands and thousands of years, storytellers have dueled with the sun. Modern scientists have further probed the sun's nature, trying to analyze it, to measure it, to weigh , and to tap its power. They also reckon with its life and death.

Hodgetts+Fung were asked to create an installation to celebrate solar power at the Frank Gehry-designed headquarters of EMR, a leading power provider for Central Germany. The task was to engage visitors and help them connect with the nature and meaning of the sun. Hodgetts+Fung began by posing the question: "What would life be like without the sun?" Their answer, "No more daisy," became the theme for the project. The design process thus began with pitch darkness.

A blacked-out room became the site where sunlight was carefully enlisted to perform a variety of experiments. Following its own schedule, it was to be captured by a great *pince-nez* of a heliostat and drawn down into the gallery to ricochet from mirrors hung from the ceiling like bats, and from there to archaic, analogue tools that transformed its energy into something the visitor could grasp. The goal was to create a magical environment powered by the sun that would entice visitors to wrestle with both the concept and physical presence of solar energy, create cosmic sounds, and chart the waxing and waning of the seasons. An array of technologically sophisticated devices accompanied by powerful images helped visitors translate their personal experiences into a deeper understanding of the complex relationships between solar radiation, energy production, and everyday life.

Overhead, solar rays ricocheted from mirror to mirror before plummeting down through lenses and filters to activate an attic full of scientific equipment that looked as though it had been gathering dust for centuries. These devices, as if sprung fully-formed from the notebooks of an eccentric

researcher, employed gears and counterweights, photovoltaic cells, collections of antique appliances, and state-of-the-art digital technology. Passing a hand over the glowing tubes of a console cradled in a rough-hewn tower, one found that it would fill the room with cosmic sounds, or throwing an old-fashioned blade switch, one watched a meter enclosed in wire-glass register the power consumption of a modern appliance as compared to that of an early model. A wisp of smoke rising above a slowly rotating turntable signaled that the sun was focused there, and the scorched circles, broken into sporadic arcs, recorded where it had gone before.

Cause and effect, tangible and immediate, connected visitors to the concept of the sun's heat. On the other hand, massive pointers, one powered by a square meter of solar energy, the other by an person pedaling a bicycle, left visitors with an indelible appreciation of solar power. If the cyclist was athletic, the race might have been even for a time, until his stamina waned or a cloud passed in front of the sun. Either way, there was an open challenge to those waiting their turn.

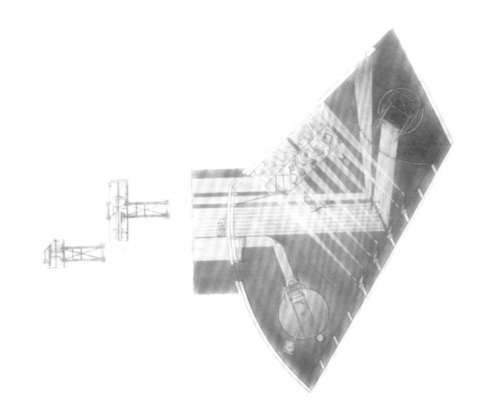

贴附在画廊外的圆形镜面把阳光聚集起来反射到画廊里，形成太阳能转换的示范

An array of circular mirrors mounted outside the gallery harvested the sunlight and directed it to the demonstrations within.

阳光被聚焦在一个低速旋转的木唱片上，留下时间的刻度，这个木唱片就是一个由机械装置带动的太阳时刻盘

An intense beam of sunlight focused on a slowly-rotating wooden disc was gradually drawn toward the center of the Solar Timetable by an elaborate mechanical apparatus.

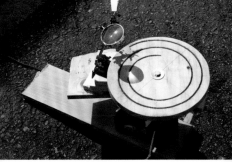

一个小型的太阳时刻盘被用来检测这个科学概念

A small-scale prototype of the Solar Timetable was built to test the concept.

对太阳能的研究和人类本身一样古老,术士、教士、预言家和魔术师是最早的研究和了解太阳性质的人,然后是苏格拉底之前的哲学家们,后来是星相学家和航海家,最后是物理学家。几千年来,讲故事的人总是和太阳有不解之缘。现代科学家更进一步地研究了太阳的性能,试图去分析它,量它的大小,称它的重量,以及利用它的能量。他们还认识到太阳的生与灭。霍德盖茨+冯被邀请设计一个装置来倡导太阳能利用,装置设在弗兰克·盖里设计的EMR总部里,EMR是德国中部最大的能量供应公司。装置的目的是引导参观者,帮助他们认识到自然和太阳的意义。霍德盖茨+冯从提出问题开始:如果没有太阳生活会是什么样子?他们的回答:不再会有雏菊了。于是构成了项目的主题。设计便从一片黑暗中开始。

现场是一个黑暗的房间,太阳光被导演成一系列不同的试验。根据它的自然过程,阳光被一个日光反射装置的巨大的眼镜片收纳,从顶棚悬挂的球拍一样的镜面反射到画廊里,然后再经一些废弃的类似装置把太阳能转换成参观者可理解的东西。装置的目的是创造一个魔幻般的太阳能环境,激发参观者理解太阳能的概念和物理特性,制造宇宙的声音,以及追踪四季变幻的规律。一排排的技术复杂的装置加上吸引人的图像能帮助参观者把他们个人体验转化到对太阳辐射、生产和日常生活等复杂关系的深入理解中去。

在人们头顶上方,阳光在镜面之间跳跃,然后笔直向下通过透镜和过滤器,激活了一个阁楼里无数的科学装置。这些装置好像摆在那里已经几个世纪了一样积满了灰尘。他们如同是从一个研究者的笔记本的图解上制造出来的,充满了齿轮和重锤,光电池,收集的古董电器,以及数字化技术。在一个加工粗糙的塔中央,参观者把手划过一个发光的控制盘时,会听到塔中充满了宇宙之声;或是打开一个老式的叶片开关,观察包在金属丝玻璃里的电表记录现代电器和早期电器用电量的不同。一缕烟从一个旋转的唱片中间冒出来,表示阳光聚焦在那里,一些烧焦的圆圈,破损成零星的弧形,记录着它们以前的形状。

因果关系,可触摸以及直截了当,使参观者理解太阳热量的概念。在另一方面,巨大的指针,一个被一平方米的太阳能驱使,另一个连接到一个人踏的脚踏车上,这种对比带给人对太阳能力量的难忘印象。如果登脚踏车的人是一个运动员的话,这个比赛可能在一段时间是对等的,直到运动员的体力下降了或是有云遮住了太阳。不管哪种情形,每个人都有机会去试试。

控制这个太阳能装置的是一个彩色发光管键盘,它具有触发功能,可以让参观者听到宇宙的声响

The glowing, rainbow-hued keyboard for the Solar Organ conducts sunlight to an array of photo-sensitive triggers that allow a visitor to release a torrent of cosmic sounds.

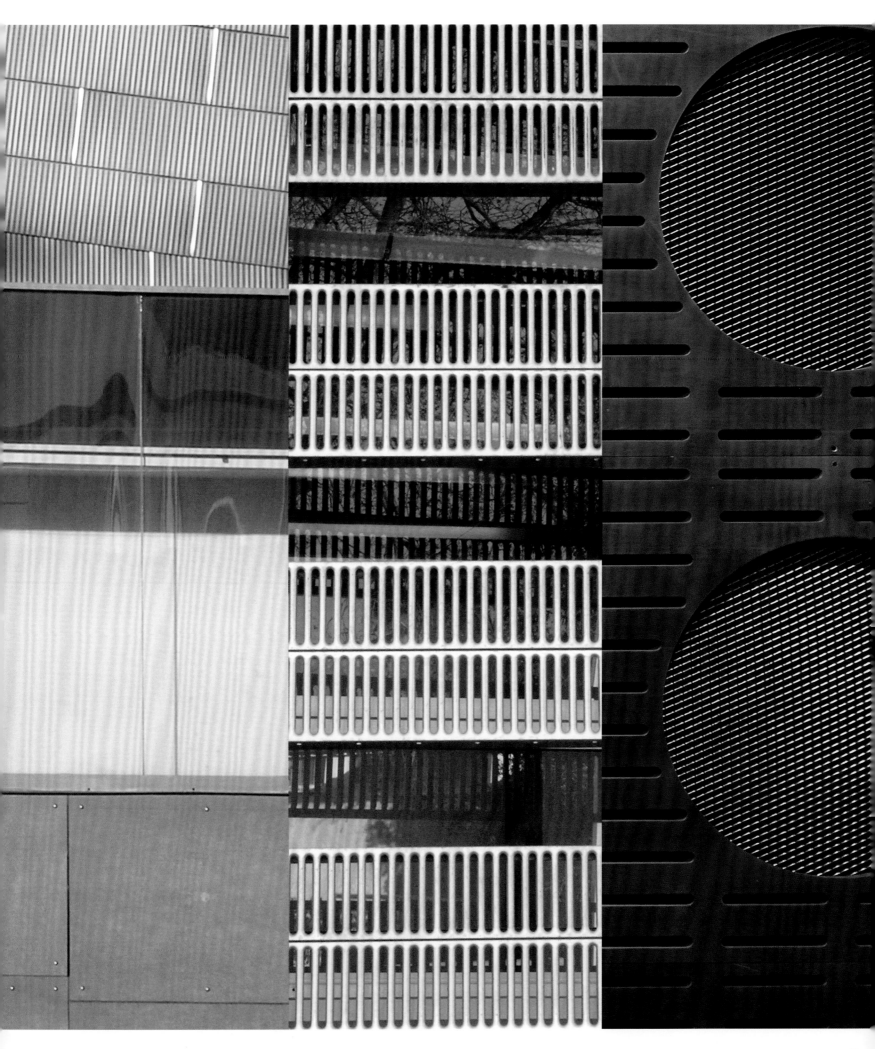

MULLIN SCULPTURE STUDIO
OCCIDENTAL COLLEGE
ECHO PARK, CALIFORNIA 1994–1997

穆林雕塑工作室
西方学院
艾柯公园大道,加利福尼亚州 1994–1997

This hard-working building sited on the periphery of the Occidental campus is as unpretentious as it is robust. Occupying a forested slope at the intersection of the college and the surrounding residential neighborhood, the Mullin Sculpture Studio was intended to serve as a gateway to the campus. In drawing together features of the site and the demands of program, **Hodgetts+Fung** developed a vocabulary appropriate to the task, "collegiate," yet not overtly institutional, so that students and faculty alike would recognize the building as a place for physical rather than cerebral activity. Conversely, it was equally important for the building to be accepted as a member of the campus community.

To accommodate the need for flexibility and functionality, the building diverges at one end. A wedge-shaped court was cut through the simple form of the structure to yield an open-air studio and exhibition courtyard. This unique "communal" space serves both to buffer and unite the building's disparate elements. The descending line described by the cut, progressing diagonally across the pitched roof, yields unexpected angles, leading the eye to a small gallery.

A pragmatic use of materials established a straightforward hierarchy: strong and impact-resistant below, where students wheel steel angles and welding equipment; fragile and light-filled above, where daylight is admitted to studio spaces. A system of grand sliding doors introduced the leitmotif of transformation, echoing the students' industry and creativity.

The rewards of such a project are to be found in the ease with which the building empowers its users. Does it demand caretaking? Is it generous? Does it overpower? Is it boring? Or does it simply perform?

Hodgetts+Fung sought to strike a balance in which the admirable qualities of a rudimentary warehouse were complimented by unobtrusive but stimulating structural and visual design elements. These were buttressed by thoughtful amenities placed "at hand"—bathrooms accessible from the courtyard; an abundance of lighting with easily-replaceable bulbs; shelving built in behind small "show" windows; ceilings high enough for second-tier storage; unobstructed circulation; and large, sliding doors to accommodate large objects and machinery.

By assuming such an inclusive attitude, **Hodgetts+Fung**'s design captures both the harmonious natural setting of this small, liberal arts college campus and the spirit and utility of a flexible building for teaching, making, and exhibiting art.

SITE MAP

穆林雕塑工作室这个利用率极高的建筑坐落在西方学院校园的边缘，看上去既结实又不矫揉造作。它占据在学院和邻近住宅区接壤的有树林的山坡上，起着作为校园入口大门的作用。根据用地的特点和任务书的要求，霍德盖茨＋冯开发了一套对项目合适的语汇，"学院制度式"而不是公然的学院式，所以学生和教师可以把这个建筑作为一个体力性的而不是脑力性的地方。相反地，这个建筑的风格被整个校园风格所接受也是一个很重要的因素。

为了实现可塑性和功能性的需要，建筑在另一端分成两叉。一个楔形的庭院切入简单的结构形状，产生了一个室外工作室和一个展览庭院。这个独特的"共同性"的空间既作为分隔又连接了建筑中截然不同的部分。庭院切入所产生的下斜线，成对角线地滑过坡屋顶，生成了意想不到的角度，把视线引导向一个小画廊。

对材料的实事求是地利用产生了一个直截了当的等级制度：底层用坚固耐磨的材料，学生在那里轮压角铁和焊接设备；顶层用脆弱的、轻质的材料，自然光可以进入工作室。一个巨大的推拉门系统引出了"变形记"的主题音乐。反映了学生的制造力和创造性。

这种项目带来的好处很容易从它的使用者身上发现。它需不需要管理？空间是不是够大？是不是太压抑了？是不是太单调了？或者，它使用起来是否容易？

霍德盖茨＋冯试图展现一种平衡，那就是一个原始的仓储建筑的令人向往的质量，被无阻碍又有启发性的结构和视觉设计元素所补偿。一些考虑周到的细节也支持着这一平衡，比如说卫生间直接和庭院连接；充足的照明都用了很容易更换的灯泡；小"展示窗"后面的固定储物架；顶棚高到可以建起夹层储藏间；无障碍的人流；以及可以运输大型的雕塑或设备的巨大的推拉门。

霍德盖茨＋冯的设计既抓住了这个小型的人文学院的和谐自然的特性，又反映了一个灵活的建筑所展示的教学，制造和艺术展览的精髓和功用。

THE WORLD OF CHARLES AND RAY EAMES
VITRA DESIGN MUSEUM, WEIL-AM-RHEIN, GERMANY 1994-1997
LIBRARY OF CONGRESS, WASHINGTON, DC 1999
LOS ANGELES COUNTY MUSEUM OF ART, LOS ANGELES 2000

查尔斯和雷·埃姆斯的世界展览
VITRA设计博物馆,WEIL-AM-RHEIN,德国 1994~1997
国会图书馆,华盛顿特区,1999
洛杉矶县立艺术博物馆,洛杉矶 2000

The World of Charles and Ray Eames, a traveling exhibition organized by the Library of Congress in conjunction with the Vitra Design Museum, was conceived as a multifaceted excursion through their life as designers, filmmakers, inventors, and collectors. The creative bustle of their office, with its myriad prototypes, experimental equipment, and trend-setting graphics, suggested an approach to the exhibition that juxtaposed conceptual sketches, evolutionary models, and finished products with the collections of folk art, whimsical graphics, and polemic writings that were the hallmarks of the Eames' working method.

Drawing the inevitable parallels with the specialized nature of many of today's design offices, **Hodgetts+Fung**'s interest was to communicate the radical nature of the Eames' method, in the hope that it would spur contemporary designers to think beyond conventional solutions. The physical components of the installation underscored the Eamesian value of constraint by the creation of a single, straightforward system of frames and infill panels, assembled to present each aspect of the designers' legacy. Reconfigured to suit the various exhibition venues, the "kit of parts" was supplemented by complementary elements designed to adapt to the specific dimensions of each space.

A single axis became the organizing principle for this installation presenting the work of nearly fifty years, beginning with the "Kazam!" machine, which the Eames installed in their bedroom in order to experiment with bent plywood, and ending with the spinning chamber they devised for an exhibition of their furniture at the Museum of Modern Art in 1946 to test the durability of their designs. Along this itinerary, visitors walked on projected soap suds from the movie "Blacktop," opened drawers still full to the brim with colorful ephemera collected by Ray Eames, and eavesdropped as documentary video portrayed friends and staff describing the experience of working with two of the last century's most celebrated designers. Such juxtapositions echoed those typical of the activities within the Eames office, providing insights to the way in which multiple perspectives influenced not only the visual aspect of the Eames' designs, but also the thinking behind the designs.

尔斯和雷·埃姆斯的世界，一个由国会图书馆和 Vitra设计博物馆共同组织的旅行展览，起初设想是一个多方面的，反映他们作为设计师、电影制片人、发明家和收藏家的生活的旅行展。他们忙碌的富有创意的工作室，其中的大数量的模型，实验设施和引导潮流的图形，都暗示着这个展览设计应当倾向于把他们的概念草图、过程模型和完成的产品并列布置，同时还有民间艺术收藏，古怪的图形和辩证的文字，这都是埃姆斯的工作方法的典型。

虽然不可避免地重复一些当今办公室设计的特殊本质，霍德盖茨＋冯的兴趣在于和埃姆斯的方法相沟通，希望启发当今设计师超越常规的想像。这个装置的物质组成强调了埃姆斯式的限制思想，建造了一个直截了当的、单一的结构框架，其中填充的展板组合在一起展示设计师每一方面的遗产。为了不同的展览会场所重新组装的"小部件组合"作为补充，用牵带的元素设计，用作每个空间中特殊尺寸的地方。

一个独立的轴线成为这个展览五十年作品装置的组织原则，从埃姆斯装在他们卧室里的为试验三合板弯度的"Kazam!"机器开始，到他们在1946年纽约现代艺术博物馆展览他们自己设计的家具所用的旋转房间，那是为了测试他们的设计作品的坚韧性。沿着这条时间旅程，参观者行走在从电影《黑色上衣》援引的突出的肥皂泡上，打开满是雷·埃姆斯收集的彩色海报的抽屉，耳旁飘过纪录片里朋友和雇员们回忆与这两位举世艺术家的工作经历。这样的罗列布置反映了埃姆斯工作室里典型的行为，从内部的观察表述出多重的视角不仅影响到埃姆斯的设计，还影响到他们设计之外的想法。

其中最有趣的莫过于霍德盖茨＋冯在为这次展览所做的研究和准备的过程中发现了DCF椅子的原型，这种椅子自1949年以来就没有被人发现过。查尔斯·埃姆斯曾经让一个车体制造商为他制造两把玻璃纤维椅子，他亲自送去了椅子的石膏模型，但是在取回它们时，他只有足够支付一把椅子的25美金。因此其中一把椅子被留在了那里，在一个金属罐的背后静静躺了近50年，直到霍德盖茨＋冯发现了它。

Of particular interest was the first prototype of the DCF Chair, which had lain undiscovered since 1949 until discovered in fabricator's workshop by H+F during their research and preparation for the exhibition. Charles Eames, having contracted with a maker of kits for car bodies to produce two fibreglass chairs from a plaster mold he delivered personally, had only $25—enough cash to pay for one of them—when he returned to pick them up. The chair he left behind was propped up as H+F found it, on a metal ashcan, for nearly fifty years.

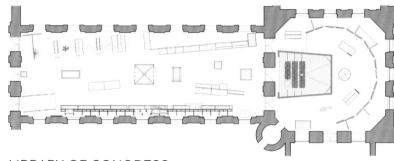

LIBRARY OF CONGRESS

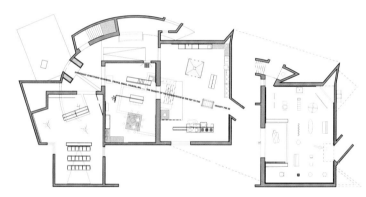

VITRA DESIGN MUSEUM

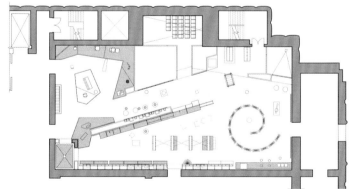

LOS ANGELES COUNTY MUSEUM OF ART

DEVELOPED FOR AN INVITATIONAL EXHIBITION ON THE
SIONED A NOVEL URBAN ENVIRONMENT AS A PROJE
POLITICO-CULTURAL TRENDS. A CITY LIBERATED FROM P
CIAL TRANSACTIONS, FROM THE NUISANCE OF REGULAT
ENA LIKE E-BAY AND CONFIGURED BY THE INVISIBLE BO
WEAL. AS A MONTAGE OF HIGHLY-DIFFERENTIATED ENCLA
DENTS, IN THESE MANGA-STYLE RENDERINGS, THE PRES
FINALLY DIGESTED BY THE APPETITES OF ITS INHABITANTS

THE RESULTING URBAN FRAMEWORK, DEVOID OF THE
SPONDS TO THE IMPERATIVES OF COLLECTIVE EXPRES
PHONES AND NEARLY INFINITE BANDWIDTHS CONSPIRE
AND CONVENIENCE ARE LESSER CRITERIA THAN IDEOLO
PHASIZE DIFFERENCES, RESULTING IN A PATCHWORK OF
CORPORATED, MOSTLY RESIDUAL ZONES. JOINED BY SW
CLAVE IS EMBEDDED IN THE MEGALOPOLITAN REGION, Y

ARCHITECTURE AND URBANISM FOLLOW SUIT, ADDING FU
HAD BEEN LIMITED TO CLASSICAL, AVANT-GARDE, AND P
TIES SUCH AS PERSONAL TRANSPORT, NOVEL LEARNING
FUNCTIONS, A NEW AND UNEXPECTED ARCHITECTURE
ENCLAVE IN SEARCH OF NEW TASTE SENSATIONS. LU
DRESSERS CROSS-POLLINATE TO A FRENZIED PITCH. AID
TION, JOBS AND CIRCULATION ARE SURGICALLY SEPARAT
E-COMMERCE. "MOM AND POP" STORES, EDGY START-U
ABLE TO FLOURISH IN HARMONY WITH THEIR CONTEXT.

E OF "IDENTITY AND DIFFERENCE," THE PROJECT ENVI-
 OF THE IMPACT OF CURRENT TECHNOLOGICAL AND
ING CONVENTIONS, FROM THE LOGISTICS OF COMMER-
 A CITY LIBERATED NOT BY SOLDIERS, BUT BY PHENOM-
OF LIFESTYLE RATHER THAN THE FICTION OF A COMMON
REFLECTING THE OFTEN IDIOSYNCRATIC VALUES OF RESI-
TIVE GRID OF THE AMERICAN CITY IS DISMEMBERED AND

RALIZING IMPULSES OF 19TH-CENTURY PLANNING, RE-
 WITH AN EXPONENTIAL INCREASE IN DIVERSITY. CELL
REATE A FRAMEWORK IN WHICH LOCATION, ADJACENCY,
D EMPATHY. POLITICAL BOUNDARIES ARE DRAWN TO EM-
F-GOVERNING ENTITIES FLOATING IN A MATRIX OF UNIN-
NG OFF-RAMPS AND DEAD-END CAPILLARIES, EACH EN-
MAINS AUTONOMOUS AND SELF-SUFFICIENT.

HIP-HOP, JAZZ, AND FOLK TO A DESIGN REPERTOIRE THAT
ODERN. POWERED BY A NEW GENRE OF URBAN AMENI-
CES, UNIQUE BUSINESS MODELS, AND DISPERSED CIVIC
GES. CULINARY ADVENTURERS LEAP FROM ENCLAVE TO
 LOOK FOR TRENDS IN SHOPS AND COFFEEHOUSES.
 THE NET, BUYERS FIND SELLERS REGARDLESS OF LOCA-
ND BASIC SALES TRANSACTIONS ARE EITHER BIG BOX OR
ND NICHE BOUTIQUES FILL THE VOID, AND ARE FINALLY

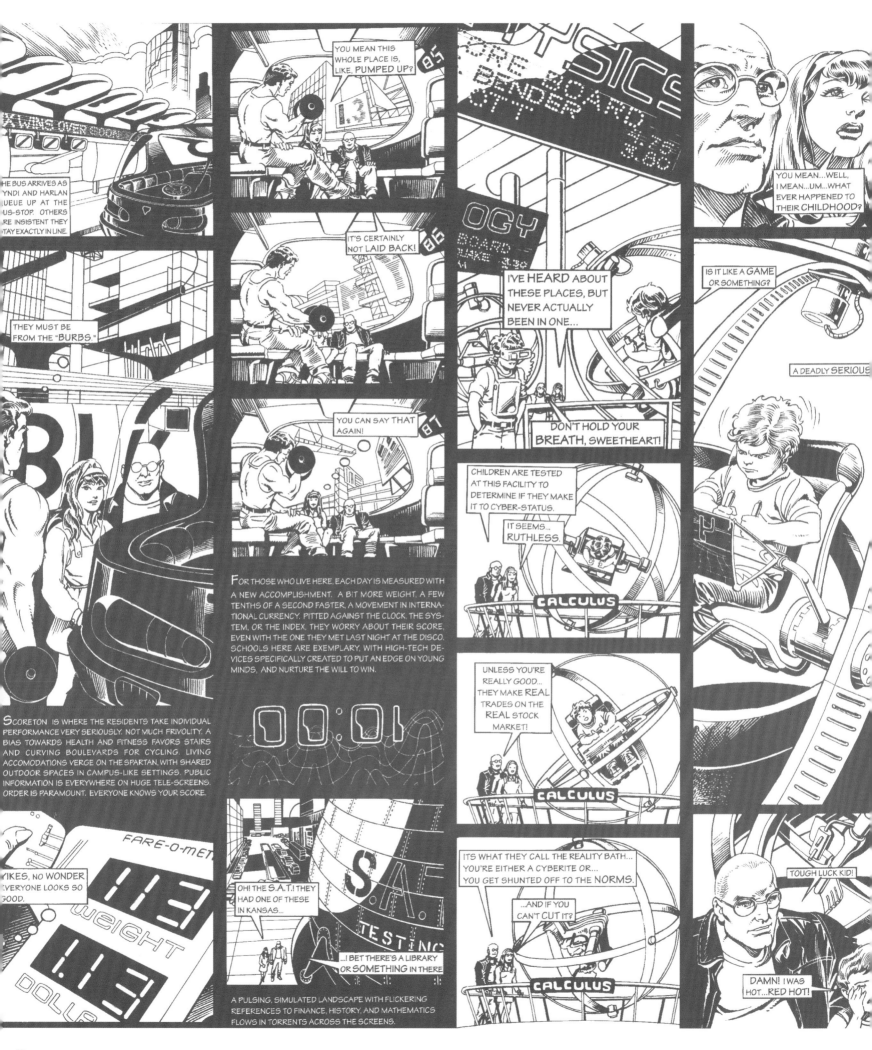

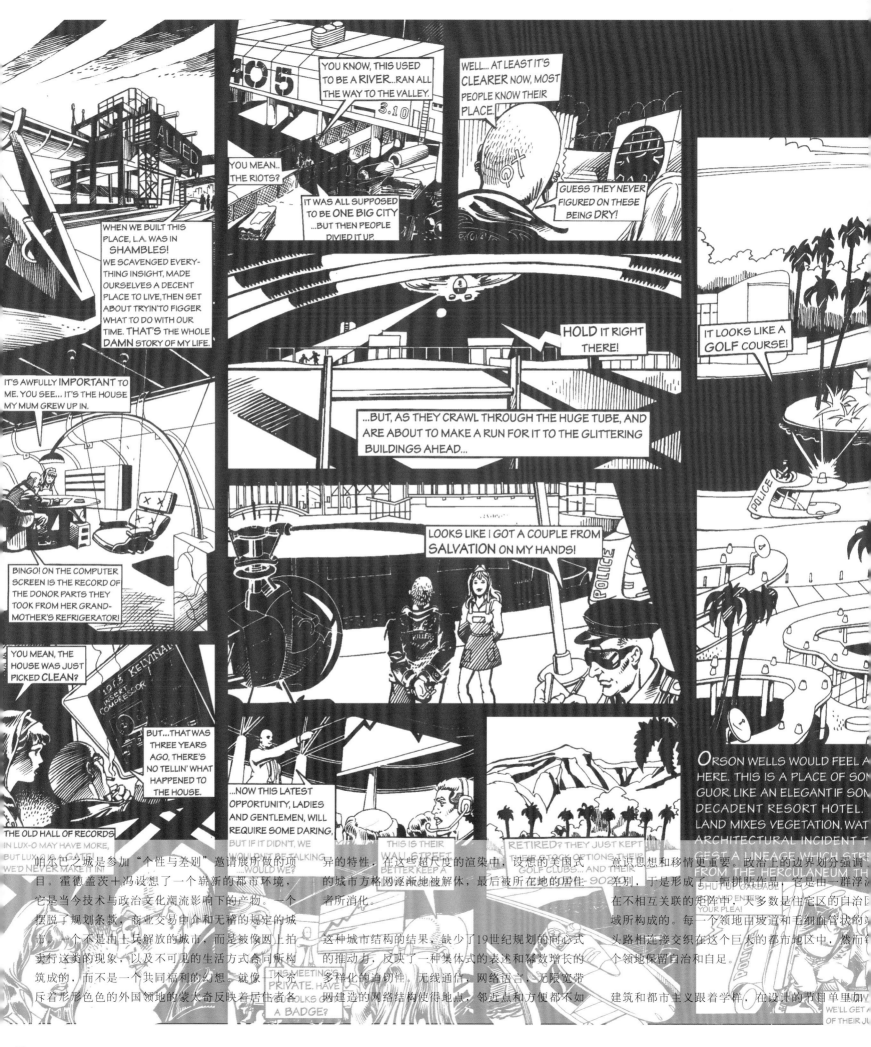

AMERICAN CINEMATHEQUE
AT THE EGYPTIAN THEATRE
LOS ANGELES, CALIFORNIA 1995–1999

美国电影礼堂
埃及剧院
洛杉矶,加利福尼亚 1995~1999

Hodgetts+Fung's renovation of Hollywood's fabled Egyptian Theatre inserted a new, free-standing structural frame into the restored envelope of the original auditorium. As the lights dim and screenings begin, high-tech midnight blue panels slide forward to produce an acoustically improved viewing experience; after the film ends, the panels retract and the opulence of the old Egyptian is visible once more. It's a larger, more wonderful version of the show at the Sinclaire pavilion's coffee bar. It's also a sly reference to the historic theatrical tradition of "atmospheric" movie houses rigged with moving scenery and other visual effects. - Reed Kroloff

霍德盖茨＋冯在好莱坞埃及剧院的改造项目中，嵌入了一种新的无支撑结构框架对老放映厅的外壳进行了翻新。当灯光暗下来，荧幕开始显现，一种高技术的向前滑行的午夜蓝板制造了一种全新的视听感受。当电影结束时，这个板子又缩了回去，老埃及的繁华得以动感再现。一个巨大的、更美妙的版本在辛克莱馆的咖啡吧里上演。这是对传统剧院利用假的动态布景和其他视觉效果来制造移动房子气氛的手法的一种聪明的借用。雷蒂·克劳夫

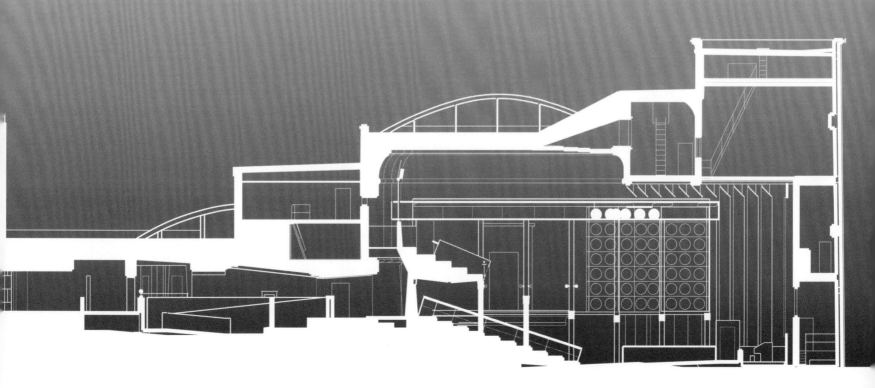

Swapping engines is a tradition in the hot rod culture of Southern California. A late-model high-performance engine can give a Depression-era coupe the performance of a Ferrari. Likewise, a high-performance cinema and black-box screening room inserted into the "classic" environment of Hollywood's Egyptian Theatre has delivered a state-of-the-art complex devoted to the art of film. Originally built in 1922 by impresario Sid Grauman, and devastated by the Northridge earthquake of 1994, the Egyptian originally consisted of a single cavernous space for performance and an outdoor lobby reached at the end of a monumental forecourt where the first Hollywood premier took place. The task was to create a radically new cinema and update the technology within the ruined shell to accommodate the Cinemathèque's programming of film and new media, while restoring the significant surviving décor and the historic courtyard inspired by ancient Egyptian architecture.

In response, **Hodgetts+Fung** devised a scheme to invert time. While the exterior was being meticulously restored to its original appearance, 700 seats were eliminated from the old theatre. A new, 616-seat cinema, the Lloyd E. Rigler theatre, was installed along with the structural armature capable of independently supporting a new balcony, air handling ducts, and a unique system of retracting acoustic panels suited to today's surround- sound technology. The more intimate, 78-seat Steven Spielberg Theatre, a black-box screening room, was sited below grade to avoid destruction of a low-slung decorative ceiling. **Hodgetts+Fung**'s solutions maintained the integrity of the original building while every aspect of the theatre's functioning—projection, sound, seating, mechanical systems, and circulation—were brought up to 21st-century standards.

Punctuated with glints from polished metal, and the unlikely decorative accents of halo-surrounded sprinkler heads, the vast interior of the theatre is an essay in contemporary opulence. The iridescent midnight-purple sheen of the acoustic panels and the moody repetition of dimpled metal create a restrained response to the original ornate sunburst of the restored ceiling. As programs begin, the

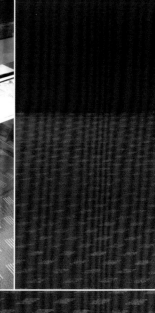

original textured sandstone walls visible through the unadorned structural armature disappear behind those panels as they extend to surround the audience with sound. Like the paintings of Gustav Klimt, the dizzying array of patterns and textures conveys a faintly "orientalizing" sensibility, which in concert with the fragments of faux-Egyptian décor produce an atmosphere at once contemporary and melancholy.

The monumental palm-studded courtyard, with its massive columns and entablature freshly restored, mimics ornamental forms traced from ancient templates. Yet beyond the entry portal, a contemporary syntax of red lacquer, frosted glass, and burnished metal prevails in the lobby. Multiple reflections scatter beneath low-slung faux-Egyptian ceilings, and angled ramps descend to the screening room submerged like a vaulted golden sarcophagus. The "magic viewing box" of the main cinema has become a transformer, in which old replaces new, and vice versa, at the push of a button.

换汽车引擎是炎热的加利福尼亚南部文化的一个传统。一个新型号，高标准的引擎可以使一个已经处生衰败期的两门车获得类似法拉利的动力。同样，一个高标准的电影院和一个放映室放置在好莱坞埃及剧院"古典"的环境中，可以奉献给电影艺术一个艺术的氛围综合体。埃及剧院最早在1922年由剧团经理人锡德·格劳曼建造，1944被北部板块地震毁坏，它曾经包含一个洞穴般的演出空间和一个室外门厅，从一个纪念性的前庭院的端头进入，在这里曾经举行过第一个好莱坞电影首映式。这个项目是创造一个崭新的电影院，在破损的外表里面更新支术以适应新的电影放映的程序和新媒体，同时要多复留存下来的装饰和受古埃及建筑启发的历史性庭院。

霍德盖茨＋冯设想了一个使时光倒流的方案。

外表被及其精细地修复成它原来的面貌，拆掉原剧场的700个座位，一个新的可容纳616人的Lloyd·E·Rigler电影院，以及电动的结构装置可以独立支撑一个新的包厢、空调和管道，一个独特的可收回的反射板系统以适应现代的环绕声技术。一个更亲密的可容纳78人的斯蒂文·斯皮尔伯格剧场(小放映厅)，设置在地下，以避免损坏低悬的装饰顶棚。霍德盖茨＋冯的设计保持了原剧场的完整性，同时剧场的每一个功能——放映、声效、座位、设备系统和人流，都提升到了21世纪的标准。

剧场室内用磨光的金属镶嵌，再加上喷淋头周围的光晕，这个巨大的室内空间设计体现了当代的丰富性。反射板那带虹彩的午夜紫的光泽，和重复的涟漪图案的金属创造了一个对原装饰母题——云缝中阳光的微妙的回应。原来的沙岩装饰墙可以从未经装饰的电动结构板的缝隙中看见，当演出开始的时候，这些结构板伸展出来把观众席包围其中，后面的沙岩就被遮挡起来。就像克利姆特的绘画，令人眩目的图案和材制的布置传达了一种淡淡的方向敏感性，和仿埃及的装饰碎片配合在一起，形成一种既现代又忧郁的气氛。

在具有纪念性的种着棕榈树的庭院里，巨大的柱子和横梁都被整修一新，根据从古代装饰线角拓下的模板设计。然而从入口门廊进去，红色喷漆、半透明玻璃和磨光的金属等现代的语汇充斥着整个门厅。多重的反射漫布在低悬的仿古埃及顶棚下面，成角度的坡道向下通往如镀金石棺一般深埋的小放映厅。主体电影院这个"魔幻西洋景"变成了一个转换器，老的换成新的，反之亦然，都在一个按钮控制之中。

MICROSOFT PAVILION
ELECTRONIC ENTERTAINMENT EXPO
LOS ANGELES, CALIFORNIA 1996

微软馆
电子娱乐用品博览会
洛杉矶，加利福尼亚 1996

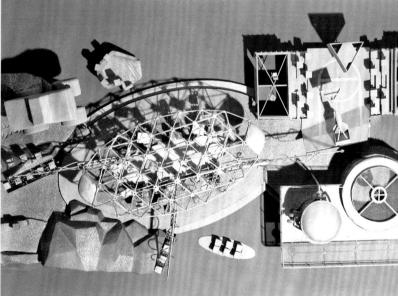

Once a year, the computer gaming industry gathers to introduce new products at the Electronic Entertainment Expo, better known as E3. **Hodgetts+Fung**'s introduction to the world of computer gaming came when the office was invited to conceive and design the environment for Microsoft's debut pavilion at E3.

This was to be a "stealth" pavilion with few identifying graphics and none of the customary costumery. Instead, attendees were immersed in individually designed "enclaves" rich with references to the games themselves: a monster truck mounted on hydraulic pistons followed the movements on a player's game-screen, while nearby, a specially-constructed elevator lowered players from a boarding platform to a circular "undersea station." To distinguish Microsoft's pavilion in the free-for-all of E3, it was imperative to animate the otherwise stationary game players, defying the common wisdom of such expositions.

Basking in the glow of the luminous screens of gaming stations, and animated by the chair lifts that raised game players to stations high above the convention floor, the gritty tubular structure called the Neural Core was a maze of reflections, laser beams, and fragmented images. Meant to suggest the dynamics of an immersive game experience, with forced feedback, high resolution graphics, and warp-speed interaction, each of the gaming stations

was enclosed by a salvaged automotive glass canopy. There, strutted out from the triangulated web of a huge, flattened ellipse, and straddled by a folding arm bearing an interactive joystick, gamers were able to grapple with the latest generation of Microsoft games.

These "enclaves" were conceived as representative of the increasingly fragmented character of the urban environment, and relished the opportunity to experiment, in a "test-tube," with the dynamics inherent in those places.

Entry to the pavilion led visitors through a winding "canyon" in a jagged steel landscape toward a glowing cloud of moving lights and images.

进入微软馆的观众将穿越一个由锯齿状钢材构成的弯曲的峡谷通道，正面的发光管制造了光线和影像的移动

Interactive computer stations on custom pedestals encouraged visitors to sample the action of a new generation of games.

交互式电脑工作站的基座鼓励参观者去尝试新时代的电脑游戏

Overhead video displays were accessed from control stations that elevated game-players to playing position.

头顶的视频显示器的进入关口都在可抬升游戏者的游戏位置的控制站上

每年一度在电子娱乐用品博览会上（通常被称为E3），各电脑游戏产业公司聚集在一起介绍新产品。霍德盖茨＋冯被应邀设计微软展馆，由此而被介绍到电脑游戏的世界里面来。

这个项目被设想成"秘密的"环境，没有可识别的图形，没有免费礼物，也没有任何惯常的商业行为。相反的是，参观者在为每个人单独设计的"领地中"领略丰富的游戏世界：一个巨大的卡车安装在一个液压活塞上，由玩游戏人在电脑屏幕上操作着；在它的旁边，一个特殊控制的电梯带着参观者从展厅地面降落到一个圆形的水下操作台。为了在完全自由发挥的E3上突出微软的形象，必须促使静止玩游戏的人动作起来，这是设计这类博览会的窍门。

在明亮的荧光屏光辉的沐浴下，玩游戏的人被椅子的电动装置抬升到远远高过地面的位置，整个这个坚韧的圆筒结构称为中枢筒，是一个充满了反射、雷射光柱和图像片断的迷宫。为了体现电子游戏体验的动感，每个游戏站都设置了反馈信号、高分辨率图像和极其高速的互动效应，并且用汽车玻璃做成的顶棚围合起来。一个巨大的椭圆形的平顶用三角形网架支撑着，游戏者骑在一个可折叠的带有游戏杆的"手臂"上，享受着最新一代的微软游戏的刺激。

我们把这些个人领地看作是当代都市环境发展的缩影，利用这个机会在这个"试管"里进行了尝试，并且给这些场所注入了内在的活力。

Each station was outfitted with a folding console containing a control stick. On activation, it would rise several meters into the air to bring visitors to the game display.

每个工作站都配置了一个折叠的操纵杆，当激活它的时候，游戏就开始了

The spine of the installation provided connections to the domains of four games. The distorted court and basket for the NBA game can be glimpsed in the background.

装置的骨架可以为4个游戏提供在线联接，从这儿可以瞥见背后的为NBA游戏设置的天井和投篮筐

In the visual overkill that characterizes the E3 convention, the unoccupied airspace above the convention floor offered a highly visible graphic opportunity. **Hodgetts+Fung**'s second design for a Microsoft pavilion used that airspace to announce an installation featuring their newest games by concealing upturned high-intensity projectors within a parade of streamlined kiosks. Projected on a narrow screen nearly two hundred feet long, a parade of text and images streamed high above the heads of convention visitors.

At the center, an elliptical capsule representing the internet zone glowed with a slowly-changing chromatic light, pulsing with luminous color and punctuated by integrated gaming stations. Designed to suggest the flux of information and interactivity provided by the Worldwide Web, the structure of each station provided links to the surface of the zone.

Encircled with a raked freeway ramp at one end, and a fog-shrouded field of medieval war machinery at the other, the clash of images and technology, from ancient to futuristic and from fantastic to realistic, was extreme yet highly disciplined. Metaphors for the action depicted in Microsoft's Gallery of Games defined each zone, creating opportunities for specific reinforcement of the salient aspects of each game.

From the ramp, the amplified sounds of a Formula One Ferrari assailed the primitive siege towers, while fallen pylons depicted the seismic destruction wrought by Microsoft's forced feedback system. Although a landscape with such seemingly violent dislocations of time, space, and content could be disorienting, in fact the opposite condition was found to be true: by exploiting the immersive quality of each quadrant, the visitor was able to navigate the whole environment with confidence.

The translucent shell of the internet gaming zone pulsed with changing colors in response to internet activity.

半透明的网络游戏区跳动的色彩随着游戏的变化而改变

MICROSOFT PAVILION
ELECTRONIC ENTERTATINMENT EXPO
ATLANTA, GEORGIA 1997

微软馆
电子娱乐用品博览会
亚特兰大，佐治亚州 1997

uthentic sounds of a Formula One Ferrari were
eatured in Microsoft's Formula One racing
ame. Broadcast from the tailpipes through
pecially configured speakers, they announced
pecial events taking place in conjunction with
e exhibition.

米拉1的失真音效是微软费米拉1游戏的重要特色，从
放器里传出的这种拉着长长的尾音的声音通告着展览
件，它也是整个展览的一部分

览中的图形来源于游戏中紧急装配，它示范了最新版
本的强权科技游戏

Graphics derived from emergency equipment
identified gaming stations designed to
demonstrate the force-feedback technology of
the latest video games.

一排流线形的架子中装置了投影仪，把微软的平面形象
动态地投射到入口的门厅上

A row of streamlined kiosks housed the
synchronized digital projectors that display
animated identity graphics above the entry.

在E3这样视觉冲击力极强的场合，地板上方的空间则是一个有良好可见性的地方。霍德盖茨＋冯给微软馆作的第二方案是利用空中装置来展示微软的最新游戏，方法是用一排流线型的小亭子隐藏的高密度投影仪，向上方投影在一条200英尺（1英尺＝0.3048米）长的狭窄屏幕上，一系列的文字和图案高高地在参观者的头顶上方闪现。

在展馆中央，一个代表着互联网的椭圆形的舱体，缓慢地变换着光谱的颜色，明亮的光彩随着节奏跳动，穿插着合成在一起的游戏站。这个设计的概念是信息和互动行为都是由互联网支持的，所以每个游戏站的结构都连接到舱体的表面。

舱体的圆形的两端，一端是倾斜的高速路坡道，另一端是雾气沼沼的战场上摆放的中世纪战争机器。

从古代到未来，从幻想到现实，这种图像和技术的冲突表现得很极端但是又高度秩序化。每一个区域的定义是微软的游戏画廊里描述动作的隐喻，这样可以为进一步加强每一个游戏的显著特点提供机会。

在跑道上，第一方程式的法拉利赛车放大的轰鸣声冲击着另一端的古代防御城堡，而城堡正在坠落的大门则描述了从微软反馈系统中提取出来的地震的情景。虽然这些看上去暴烈的，时间、地点和内容错位的场景会让人迷失方向，但事实上，被证实的常常是相反的情形：在探索每一个象限中让人沉迷的游戏之后，参观者们就可以充满信心地游览整个展区了。

Hodgetts+Fung were invited to create a second exhibition at the headquarters of EMR based on the phenomenon of wave power—a dynamic force, always in motion, and best known by the ascending and descending lines of the sine curve. Everyone has experienced a simple form of wave power—pushing a child in a swing or skipping rope. The pulses of energy we call "waves" can be found at every level of existence, from the nanosecond-long orbits of electrons to the magnetic pulses that course through deep space. They are embodied in the transmissions of our cell phones, keep our clocks on time, and heat up our frozen meals. Despite their pervasiveness, however, they are famously hard to portray.

The installation design for Wave Power demanded the invention of innumerable devices capable of engaging visitors with the fleeting nature and inherent beauty of waveforms without surrendering to technical explanations or computer simulations.

The hand, the eye, the ear, and the mind were called upon to initiate each sequence of the installation. By pulling, tuning, listening, and looking, the constant mingling and reinforcement characteristic of wave mechanics were made both legible and tangible.

Swinging in great arcs, seven golden discs grazed the air above visitors' heads, producing a chorus of sounds as sensors traced embedded codes. Wide belts descended from pivot points to station points, where a rhythmic pull accelerated each pulse. Concentric patterns overlapped one another on a carpet of light in the sound garden, near which rapt visitors could observe rippling, overlapping circles as musical notes varied from soprano to bass. Overhead, seamless nodes in a long shining tube magically multiplied, before disappearing altogether. As a demonstration of the concept of wavelength, the subdivisions became shorter as the frequency increased.

ZKT WAVE POWER
ELEKTRIZITÄTSWERK MINDEN-RAVENSBERG
BAD OEYNHAUSEN, GERMANY 1997–1998

ZKT波状能展览
新能量论坛
巴特恩豪森，德国 1997~1998

一个类似自行车轮子的复合张力结构被应用来维系光学读表盘中心点的精密度

An intricate tension structure similar to that of a bicycle wheel was employed to maintain the exacting concentricity required by the optical reading devices.

由钟摆的运动产生的波状线被数字化展现，伴随着声音的起伏而跳跃

The complex waveforms generated by the motion of the pendulums were translated to a digital display to produce a visual analogue of the resulting sound.

摆的频率被激光刻录在一个光学录音带上，它的音调
冲摆的运动速度相关

soundscape transferred to an optical sound
ck was "read" by lasers fixed to the toe of
ch pendulum, producing a pitch directly
ated to the velocity of the pendulum.

一个发光板界定了这个声音花园，当追寻声音的微波每
一次划过中心线时，使人回想起禅的意味

A luminous plane defined the sound garden,
on which ripples of sound trace ever-changing
concentric lines, recalling the raked sand of a
Zen garden.

德盖茨＋冯被邀请在EMR总部设计第二个展览，
状能现象，一个变化多端的力量，不停地运动，
常可以在上升和下落的正弦曲线上看到。每个人
有一些简单的波状能的体验——把孩子推到正在
动或跳动的跳绳里去的时候。这种我们称之为波
能量脉冲可以在每个存在的阶段里找到，从电子
亿分之一秒长的振动速度到穿越宇宙深层的电磁
冲。它们传递着我们移动电话的信号，使我们的
表准时，以及溶解冷冻的肉。除去它们无所不在
这种特点，它们也是著名地难以描绘。

个波状能的装置设计需要发明许许多多设备，可
使参观者领略自然现象和波状能的魅力而不用作
多的技术性解释和电脑模拟。在每一个步骤中：
、眼、耳和脑都被调动起来。通过拉、转、听和

看，波状能设备的不断混合和加强的特性变得既可
见又可触摸。

七个金色的圆盘轻飘飘地悬挂在参观者头顶上方，
摆成一个大弧形，当感应器解读已存好的密码时，
它们能产生悦耳的和声。一个宽皮带随着有节奏地
拉伸并依次增量，从中枢点转化到静止点。在声音
花园里，向心形的图案构成了一个光的地毯，当音
乐从高音降到低音的时候，全神贯注的参观者可以
观察地上变幻的涟漪形重叠的圆圈图案。在头顶上
方一个长长的、发光的管子中，无缝隙的波点神秘
地重复再生，直到一起消失为止。作为一个波长概
念的演示，当频率增加的时候，分解再生的速度也
变快了。

Eight massive media towers arc through the five-story atrium of this shopping complex near Tiananmen Square in Beijing. Eye-to-eye with shoppers on the fourth level, the towers direct attention downward to the ground floor, where a gigantic graphic depicts the spotlights, mountain ranges, and surfboards associated with Hollywood, home of Universal Studios.

This is Universal Experience, an entertainment center designed to introduce Western entertainment concepts to the Chinese market. From gated "experiences" offering up samples of full-scale theme park rides such as "Jurassic Park," to cascades of automated follow spots, this convergence of motion pictures, merchandising, and food service is a unique experiment in global marketing.

Surrounded by international brands, and serving as a thoroughfare for those exiting the hub of Beijing's extensive subway system, the complex is a dynamic, three-dimensional space aglow with the light of giant neon signs recalling both surf shops and the car culture of Los Angeles. A retail store laced with references to Hollywood films, a pan-Asian restaurant, and a unique photo studio equipped with bluescreen superimposition gear protrude into the atrium to create a colorful gathering of unique shapes that interacts graphically with the plane of the floor.

这是一个位于天安门附近的购物中心，八个大体量的信息塔弧形地围绕着5层高的中庭。在第四层上，面向着购物者的塔把人们的视线向下引向底层地面，聚焦的中心是一个巨大的图像，带着山脉和滑板的好莱坞环球电影厂。

这就是环球体验，为中国市场引进西方娱乐概念而设计的一个娱乐中心。从原尺度主题公园的范例，比如说"侏罗纪公园"，到自动的音乐瀑布，这个汇合了电影、商业和餐饮的娱乐中心对全球市场来讲是一个独特的试验。

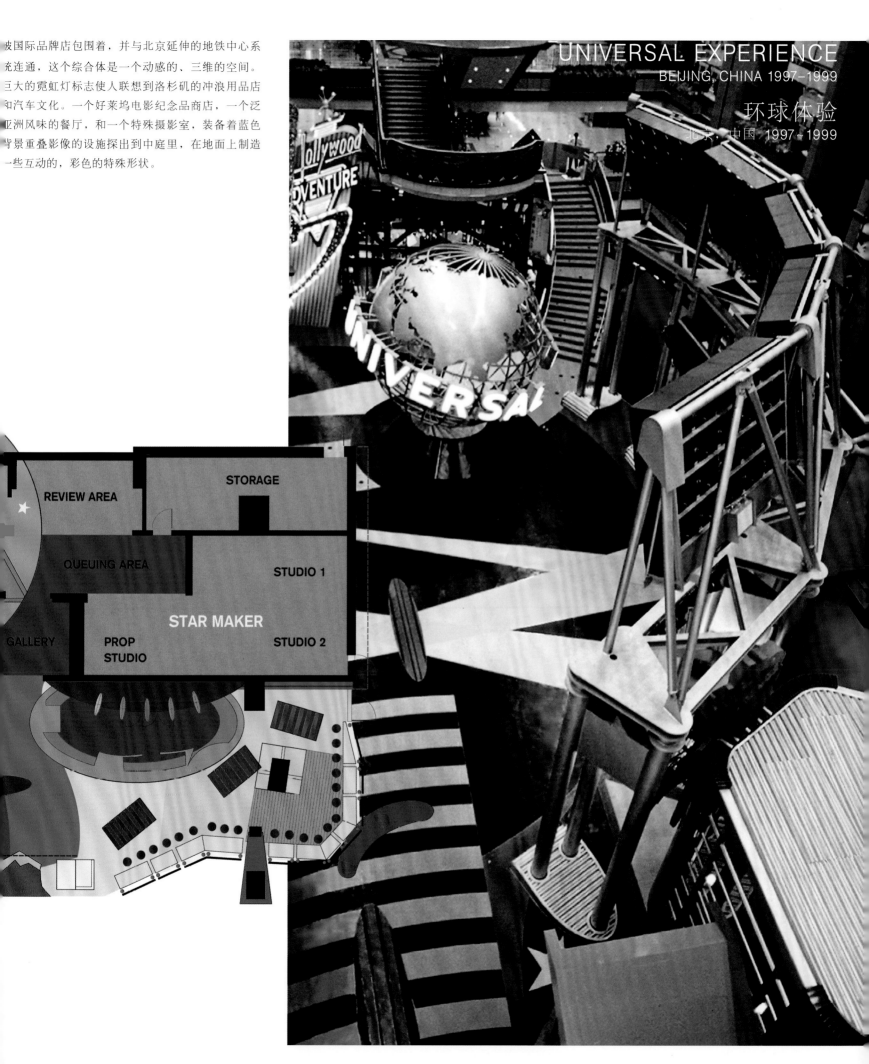

被国际品牌店包围着，并与北京延伸的地铁中心系充连通，这个综合体是一个动感的、三维的空间。巨大的霓虹灯标志使人联想到洛杉矶的冲浪用品店和汽车文化。一个好莱坞电影纪念品商店，一个泛亚洲风味的餐厅，和一个特殊摄影室，装备着蓝色背景重叠影像的设施探出到中庭里，在地面上制造一些互动的，彩色的特殊形状。

UNIVERSAL EXPERIENCE
BEIJING, CHINA 1997-1999

环球体验
北京，中国 1997-1999

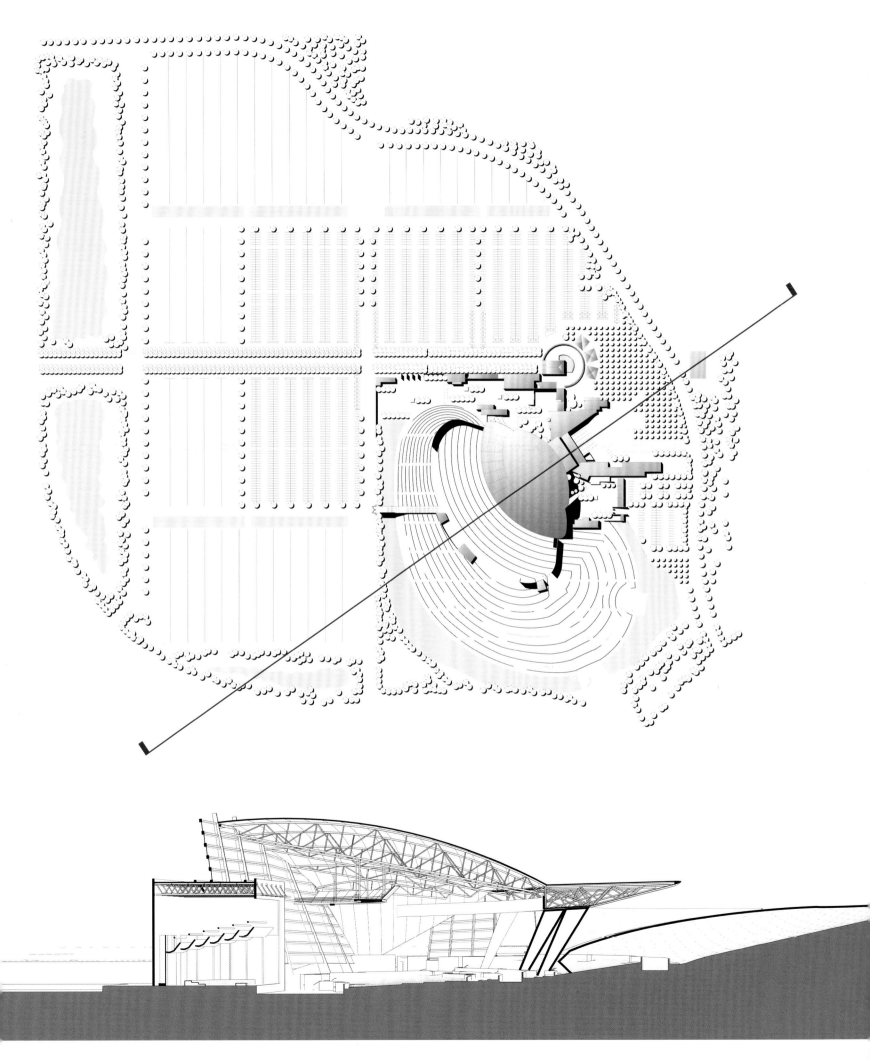

BROOKLYN PARK AMPHITHEATRE
MINNESOTA ORCHESTRAL ASSOCIATION
MINNESOTA 1999–2000

布鲁克林公园露天剧场
明尼苏达交响协会
明尼苏达 1999~2000

As perhaps the last vestige of a culture in which effective mass communication required the physical presence of a significant proportion of society, the modern amphitheater exists outside of the warp-speed web of communications we have come to take for granted. Real time, physical presence, and tangible volume combine to create an unmediated experience, which is becoming far less common as the technological is fetishized in our time.

Set against the endless horizon of the Minnesota plains and nestled into the hollow of a gigantic man made earth form, the burnished enclosure of the Brooklyn Park Amphitheatre spans nearly four hundred feet to provide shelter for an audience of 8,000. This is a public space, designed in light of urban principles rather than theatrical conventions. Like the Greek stoa that preceded it by more than two millennia, there are spaces for vendors, for taking refreshment, and for sanitation, all housed in a layered, low-lying complex. The ceremony of entrance, the sense of shared experience, and the promotion of common values were all factored into Hodgetts+Fung's design, which was intended to maximize the amphitheatre's effect upon the cultural life of this community.

The theatre is approached along a descending, tree-lined promenade that gathers audience members from their parked vehicles, channeling them past an encircling pond before admitting them to a ticketing plaza dominated by the glowing profiles of faceted sound walls. Inside, the seating is arranged on multiple terraces to condense audience energy even at sparsely-attended events. The design of such a show-building complex has become particularly challenging as the competing needs of an increasingly diverse array of presenters, promoters, and media technologies requires stages to undergo constant transformation—often with turn-arounds of less than twenty-four hours.

The earth form made of the soil removed for the construction of the amphitheater's 10,000 parking spaces makes reference to native American burial mounds common in this part of the United States. Rounded rather than angular, and offering a multi-plicity of grades owing to the elliptical plan, the hill stands in sharp contrast to the surrounding topography, yet modulates the natural landscape in such a way that the amphitheater seems inevitable. A spiraling ramp connects large-scale outcroppings for events that interrupt a grand stair at regular intervals.

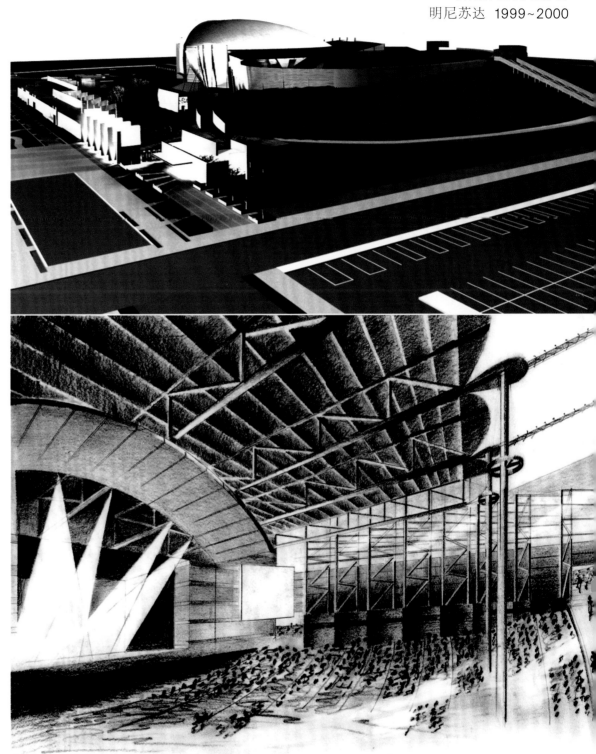

也许是一个文化的最后痕迹,大众的交流还需要一个实际意义上很大比例的社会团体,现代的演剧场在我们已经很依赖的网上信息交流之外所存在着。真实的时间,真人的参与,可触摸的空间,合在一起创造了一种没有中介者的体验,这在当今技术发展的时代已经越来越不普遍了。

以一望无际的明尼苏达平原为背景,倚居在巨大的人工堆砌的景观的空隙中,表面光亮的布鲁克林公园演剧场横跨近四百英尺(1英尺=0.3048米),形成了一个可容纳8 000听众的娱乐设施。这是一个公共场所,以城市设计的原理设计而不是剧场的常规。就像历经两千多年的古希腊的stoa(商业建筑)一样,这里有小铺子、小吃点、卫生设施,全集中在一个排列起来低层的综合体中。显眼的入口,感官分享的体验和共同价值的提升,都反映在霍德盖茨＋冯的设计中,这个设计的主旨是极大化地形成演剧场对当地社区文化的影响。

剧场是从一个下坡的林荫道引入,听众从停车场汇集到林荫道,沿着道路经过一个圆形的水池才进入围合着声光墙的售票广场。在内部,座位布置在不同的平台层上,这样一来能即使在人数最少的时候也能凝聚观众的能量。设计这样一个娱乐性建筑综合体是一个特别的挑战,不断增加,多样化的演出团体,宣传人和媒体技术都要求舞台设计经常性地改动——经常在24小时内全盘改变。

我们把建设演剧场的10 000个车位所挖掘的土壤堆砌成形,参考了当地原始印地安人墓穴的形式。因为是椭圆形的平面,堆砌的小山采用圆角而不是棱角的,并且有着多重的坡度,这与周围的地理环境形成鲜明的对比,然而却在另一种形式上重复了自然景观的规律,使得演剧场看上去像是必然形成的。一个旋转的坡道连接着举办大型活动的露台,这些露台每隔一定距离交接到大楼梯上。

THE HOLLYWOOD BOWL
HOLLYWOOD, CALIFORNIA 1999–2004

好莱坞碗形剧场
好莱坞，加利福尼亚州 1999~2004

自从1923年开始使用以来，为解决其内部功能和形式的矛盾，好莱坞碗形剧场共经历了八次主要的改造。记忆和声学合在一起使设计形成了一个新的躯壳。根据1928年的改造，声学设计不实用而形式设计则是汲取了百万个资助人的记忆。霍德盖茨+冯参考了这个时期的设计，在原来的几何形式的基础上拓展三维的想像。他们的宗旨是寻求这种平衡，即运用21世纪声学技术，并且考虑到人们喜爱的场所精神和好莱坞早期建立的品牌形象。

他们设计的中心是发明一个新的声学装置，使得几乎看不见的设施能提供一种前所未有的室外剧场声效控制。这个设施包括一个60/90英尺的铝和玻璃钢折叠的椭圆和半透明的反射板，它们形成一个可调节的反射面，反射表演者之间的声音，使其达到音率的平衡。这个飘浮的虹彩般的形状悬挂在新改造的弧形拱下，静止地飘浮在观众的头上，同时具有安静和几何张力的特性。

半透明板被设计成一个简单的可允许几乎无数范围调节的屏幕，它们在声学乐器表演的时候几乎是放平的，要想放大或电子化表演时，只需要将它们"折叠"成竖直的形状。当用于旅行演出时，这些板子可以用电动绞车转换成存储的状态。

虹彩装置和水平线成十度角，包围在麦龙和亨特设计的室外剧场之上，它和观众席很好地结合着，使表演者和18 000观众的视觉关系更加直接。以好莱坞山为背景看过去，这个新的、巨大的拱看不出一丝其中包含有复杂的技术设施的痕迹，让人们可以简单地欣赏被提高了的音乐质量。

Since its inauguration in 1923, the Hollywood Bowl has undergone a total of eight major efforts to resolve the inherent contradictions between form and function that have bedeviled it. Memory and acoustics acted together to shape the design for a new shell. Referencing the 1928-era icon whose acoustics were inherently impractical but whose shape is embedded in the memories of millions of patrons, the new shell designed by Hodgetts+Fung is based on a three-dimensional exploration of the geometry underlying the original. Hodgetts+Fung's intent was to strike a balance between bringing the Bowl's acoustical and theatrical technology into the 21st century, and considering the spirit of the much loved place and the "brand" image established in the Hollywood's early years.

The crux of their project was the invention of a new type of acoustic device in which a largely "invisible" technology would provide an unprecedented level of sonic control for the outdoor amphitheater. Composed of a sixty-by-ninety-foot aluminum and fiberglass ellipse spanned by folding, translucent panels, the acoustic device forms an adjustable reflecting surface that disperses sound among the performers, enabling them to achieve a musical balance. The floating, halo-like form hovers beneath the curving arch of the newly-engineered shell, at once in repose and charged with geometric tension as it hangs motionless above the audience.

Designed to permit a nearly-infinite range of adjustments that are made by means of a simple touch screen, the translucent panels lie almost flat during acoustical instrumental performances. Adjusting them for amplified or electronic performances is a matter of "folding" them into a vertical configuration. When deployed for a road show, the panels are flown by electric winches into docking stations between the baffles.

Inclined about ten degrees above horizontal to embrace the geometry of the Myron Hunt designed amphitheater, the halo optically engages the crest of seating, so that a performer's perceived relationship to the audience of 18,000 is far more immediate. And seen against the backdrop of lush Hollywood Hills, the new, larger shell gives no hint of the radically reconfigured technology contained within, allowing the audience to simply enjoy the enhanced musical experience.

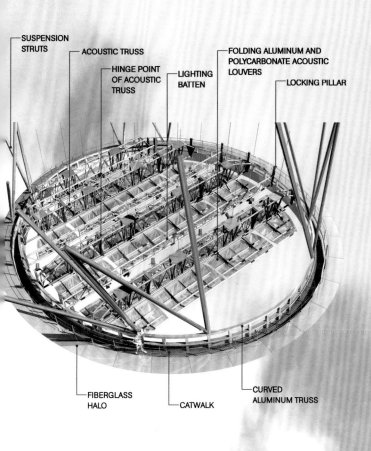
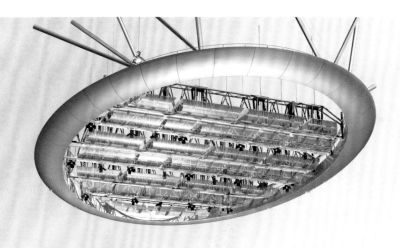

- SUSPENSION STRUTS
- ACOUSTIC TRUSS
- HINGE POINT OF ACOUSTIC TRUSS
- LIGHTING BATTEN
- FOLDING ALUMINUM AND POLYCARBONATE ACOUSTIC LOUVERS
- LOCKING PILLAR
- FIBERGLASS HALO
- CATWALK
- CURVED ALUMINUM TRUSS

Hodgetts+Fung's kinetic intervention becomes part of the theatrical experience at the Hollywood Bowl, where the architects were faced with bringing the diametrically opposing forces of nostalgia and progress into harmony. Since it was established in the early 1920s, the beloved amphitheatre in the Hollywood Hills, with its signature, segmented megaphone shape, has been plagued by dilemmas that have challenged its natural acoustical capacity. Over the years, architects including Frank Gehry have attempted to "fix" the problem, to little avail. Hodgetts+Fung's solution maintains the original Art Deco-inspired telescoping form, but they've added a new and marvelous element. A large, metal-clad, elliptical object hovers within the receding arches as if ready to sail out over the audience in a scene stolen from "The Day the Earth Stood Still." In fact, the device is an almost infinitely adjustable acoustic damper as well as an armature to support lighting and visual effects equipment. Of course it reconfigures itself at the touch of a screen, and the audience gets to take part. This is Hollywood, and Hodgetts+Fung invite everyone to be an extra.
- **Reed Kroloff**

好莱坞碗形剧场，霍德盖茨＋冯的活力已成为了这个剧场体验的一部分，建筑师面临的难题是如何协调怀旧和前行这两种反作用带来的压力。这个建于上世纪20年代早期的露天剧场为好莱坞山的居民所热爱，它标志性的圆缺传声筒造型在传声系统上的缺陷一直困扰着人们。在过去这些年里，一些建筑师包括弗兰克·盖里也曾经尝试去修补这个缺陷，但都没能最终解决问题。霍德盖茨＋冯的方案保留了这个剧院像望远镜一样的外形，他们增添了一些新的、不可思议的要素。一个巨大的包着金属外衣的椭圆形气垫形成一个可进退的弓拱，可以让观众置身于类似"The Day the Earth Stood Still"的场景中。这个装置最大限度地调节了听觉范围，并成为灯光和视效装备的支撑器械。当然它可以自我调节与屏幕的接触，观众也可以掌控一部分。这就是好莱坞，霍德盖茨＋冯带给所有人的意外。**雷蒂·克劳夫**

外壳分析
对管弦乐的声音传播需求进行全面分析后得出了新外壳的几何形态

共震研究
声学反射的椭圆形态来源于Myron Hunt设计的环形座位区

折叠天窗
其中一种声学反射系统是通过舞台上方的顶棚板将声音收集起来

SHELL ANALYSIS

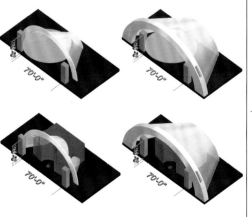

A thorough analysis of orchestral requirements versus those of touring shows led to the geometry of the new shell.

EPICENTER STUDY

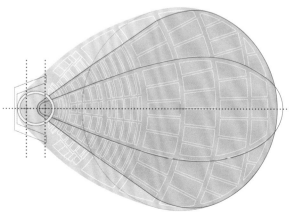

The elliptical form of the acoustical reflector was derived from the geometry of the Myron Hunt-designed seating area.

FOLDING LOUVER

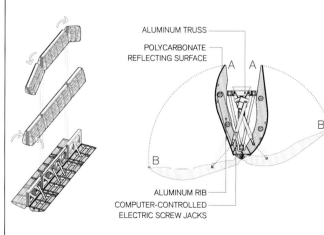

A one-of-a-kind acoustic reflector system achieves acoustic sophistication yet can fold in two planes for storage above the stage.

灯光槽
钢制骨架支撑着灯光槽，电线和开关被有序地植入其中，并可抵御风和振动

LIGHTING BATTENS

trusses support lighting battens, span the halo, into channels designed for seismic and wind

电脑控制
电脑控制的声学反射系统将依据声音环境调节音量的大小

COMPUTER CONTROL

Computer-activated reflectors will enable the acoustic environment to adapt quickly to large and small ensembles or reinforced sound.

喷水截断
一个独立的铝制悬空结构支撑着一个轻盈的房梁和玻璃纤维滑轮

WATER-JET CUT

An independent aluminum substructure supports a lighting cat-walk and the fiberglass halo element.

锁控策略
线性装备固定了可移动的天窗和灯光槽，并可抗御风和地震

LOCKING DEVICE

Teflon-lined bayonet fittings stabilize the movable louvers and lighting battens against wind and seismic forces.

After twenty-five years in Craig Ellwood's elegant Miesian extrusion, the students at Art Center College of Design petitioned their new president, Richard Koshalek, for some R&R. It should, they suggested, come in the form of a building meant for hanging out, where students could unwind after hours behind their iMacs. It should, they suggested, be a counterpoint to the disciplined environment within which they wrestled with Maya and related peripherals. It should, they hoped, be a place where anything goes—even smoking.

Jutting over a switchback ramp, sheltering a coffee bar and student gallery, the Sinclaire Pavilion is a "lounge" designed for socializing, where students can also stage spontaneous events and presentations. Situated at the apex of a knoll overlooking the clipped green lawn surrounding Ellwood's building, the student lounge is the only other freestanding structure on the sixty-acre site. Diminutive—only about 2,000 square feet—in comparison to the main building's 120,000 square feet—and resolutely *au natural*, the perfect foil for the highly-conditioned spaces of the other building, the lounge is the antithesis of the pristine orthogonals of Ellwood's design. Angled struts and complex junctions dramatize the structure. Slabs of steel form railings. An almost primitive vocabulary of galvanized steel, cement board, and corrugated aluminum siding is bolted, field-welded, and troweled into shape.

Hodgetts+Fung's aim was to create an environment that celebrated the virtues of mechanics, banishing for a time the mechanical systems, motors, and digital displays we've all become used to, and relying on human power alone. A brace of kinetic elements pivot, swivel, and glide, emphasizing the roots of Art Center's industrial design curriculum. Levers and cranks open up a giant window, while a pivoting wall hugs the curve of the room to ward off the deer who often visit the site. At the center of the space is a cubic volume. A counterweighted panel folds around its corner to pirouette gracefully upwards, signaling with an orange lining that the espresso bar concealed within is open for business.

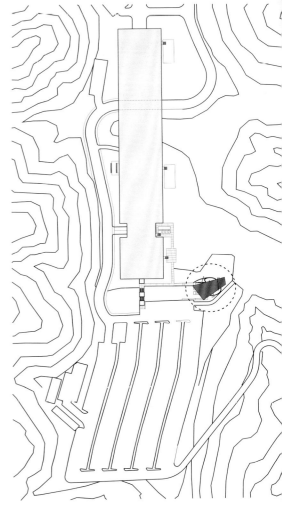

SINCLAIRE PAVILION
ART CENTER COLLEGE OF DESIGN
PASADENA, CALIFORNIA 2000-2001

辛克莱馆
艺术中心设计学院
帕萨迪纳，加利福尼亚州 2000-2001

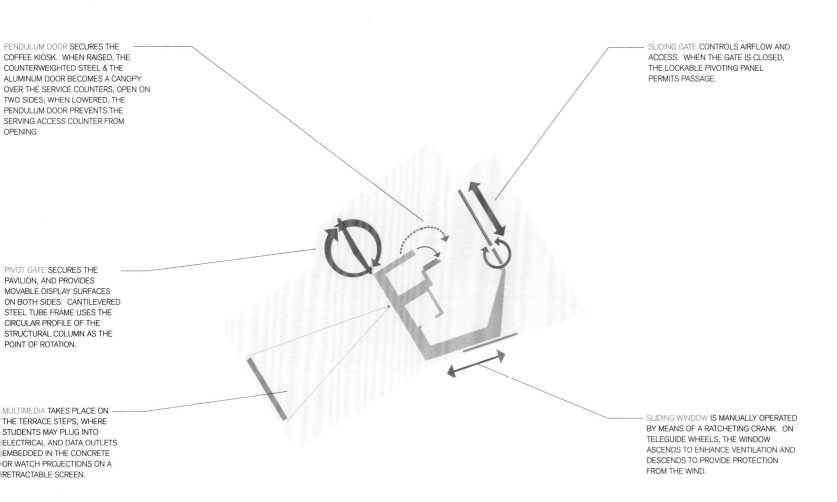

PENDULUM DOOR SECURES THE COFFEE KIOSK. WHEN RAISED, THE COUNTERWEIGHTED STEEL & THE ALUMINUM DOOR BECOMES A CANOPY OVER THE SERVICE COUNTERS, OPEN ON TWO SIDES; WHEN LOWERED, THE PENDULUM DOOR PREVENTS THE SERVING ACCESS COUNTER FROM OPENING

SLIDING GATE CONTROLS AIRFLOW AND ACCESS. WHEN THE GATE IS CLOSED, THE LOCKABLE PIVOTING PANEL PERMITS PASSAGE.

PIVOT GATE SECURES THE PAVILION, AND PROVIDES MOVABLE DISPLAY SURFACES ON BOTH SIDES. CANTILEVERED STEEL TUBE FRAME USES THE CIRCULAR PROFILE OF THE STRUCTURAL COLUMN AS THE POINT OF ROTATION.

MULTIMEDIA TAKES PLACE ON THE TERRACE STEPS, WHERE STUDENTS MAY PLUG INTO ELECTRICAL AND DATA OUTLETS EMBEDDED IN THE CONCRETE OR WATCH PROJECTIONS ON A RETRACTABLE SCREEN.

SLIDING WINDOW IS MANUALLY OPERATED BY MEANS OF A RATCHETING CRANK. ON TELEGUIDE WHEELS, THE WINDOW ASCENDS TO ENHANCE VENTILATION AND DESCENDS TO PROVIDE PROTECTION FROM THE WIND.

1. EXISTING STAIRS & RAMPS
2. ENTRY / SLIDING GATE
3. BENCH
4. FOOD SERVICE / PENDULUM DOOR
5. RESTROOM
6. TERRACE STEPS
7. ACCESS TO OUTSIDE
8. PATIO
9. VENDING MACHINES
10. A/V STORAGE BELOW
11. SLIDING WINDOW ABOVE
12. STUDENT EXHIBITION WALL
13. PIVOT GATE

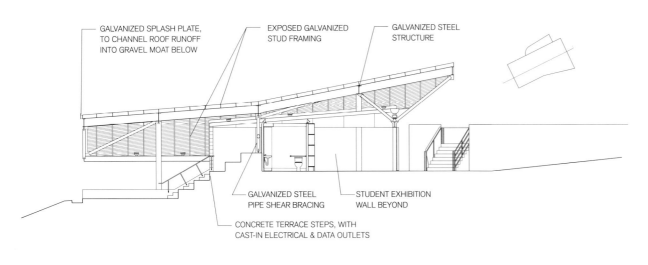

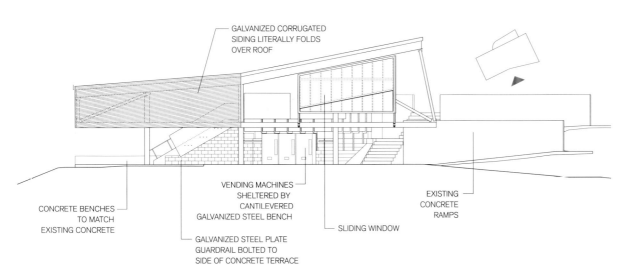

The Sinclaire pavilion is a small building sitting in the very large shadow of Craig Ellwood's masterpiece. Irregular and shed-like, the little pavilion is quirky and rough. In comparison to the elegant repose of Ellwood's, it looks restless, as if it might change shape. And it does. At the heart of the design is a mute cubic core. With the flick of a wrist, the cube comes apart, as one of the quadrants floats up like piece of industrial origami to reveal a bright orange interior. This is clever, both as a bit of architectural legerdemain and as a reference to the Art Center's renowned industrial design program. This building doesn't try to be pretty. Why would it? Nothing could be prettier than Ellwood's building. Consequently Hodgetts+Fung chose to open up a dialogue across theoretical positions. - **Reed Kroloff**

辛克莱馆是一个被Carig Ellwood的杰作的光环所笼罩的小型建筑。这个非常规的、像车棚状的小建筑好像有些粗糙。与Ellwood的优雅、统一相比，它看起来有些不安，外形容易起变化。是的，这就是这个设计的精髓所在。像一个轻轻弹开的袖口，这个立方体有些变异。就像一个浮动的四分之一圆的工业元件，里面有着明亮的橘红色里衬。这很聪明，在带有一点建筑戏法的同时，反映了艺术中心设计学院的最有影响力的专业是工业设计。这个建筑没有尝试去进行漂亮的修饰，为什么会这样呢？不可能有比Ellwood的建筑更漂亮的东西了，因此霍德盖茨＋冯选择了去开启一个穿越理论领域的对话。**雷蒂·克劳夫**

在Craig Ellwood设计的优雅的密斯型建筑里经历了25年之后，艺术中心设计学院的学生向新校长Richard Koshalek请示一个新的休闲场所，他们建议这应当是一个供学生课后放松闲谈的地方，以解除他们长时间在电脑屏幕后工作的疲劳。他们还认为这个建筑应当有一个和严谨的学习环境相反的气氛，并且希望可以绝对放松自由，包括抽烟。

辛克来馆从一个回旋的坡道上伸展出来，包括一个咖啡吧和一个学生画廊，它是一个学生可以举办自发性的活动和汇报作品的社交"沙龙"。这个学生中心坐落在一个小山丘的端部，俯瞰着修剪过的草坪围绕着Craig Ellwood的主楼，在这块6英亩（1英亩＝4 048.58m²）的用地上，它是除主楼之外惟一一个独立的建筑。相比较12万平方英尺（1平方英尺＝0.0929m²）的主楼，辛克来馆像是微缩景观，只有2 000平方英尺，然而它是完全自然的，是主楼的高度空调系统环境极好的陪衬。这个沙龙与Ellwood设计的正交的建筑正好相反，成角度的支撑构件和复杂的交接使得这个建筑富有戏剧性。用钢板制成栏杆。一组几乎原始的语汇，镀锌薄钢板、水泥板和波浪形铝制面板在现场被铆接、焊接或铲成不同形状。

我们设计的目的是为了提倡手工机械技术，在一段时间里可以废除我们所习惯的空调设备系统，电动马达和数字屏幕，而只是依靠人力。一系列的枢轴、轴承和滑轨的使用，强调了艺术中心学院的工业设计的根基。一个巨大的窗户可以由杠杆和曲柄开启，一个带转轴的墙可以沿着弧形的房间旋转挡住经常来访的野鹿。中心空间有一个立方体，转角上一个自我平衡的木板优雅地向上翻转过来，露出里面的橘红色，标志着藏在里面的咖啡吧开始营业了。

Taking their place on a heavily-traveled shopping strip in Alhambra, a dynamic Chinese enclave east of downtown Los Angeles, the calligraphic forms of this banking building are pared down to basics: a roof, an enclosure, a point of entry, and a sign. Executed in aluminum, glass, and Portland cement plaster, each form is elemental, governed by primary geometry and uncompromised by allusions to utility or brand.

Changing perspectives as one drives by animate the building's relationship with cruising cars. A roadside graphic sharing the material palette of the building echoes the graphic atop two exposed-steel columns, and the westward tilt of the roof seems to flatten as it floats towards the streaming traffic. The three-dimensional construction yields a succession of views governed by parallax and speed.

Support for the roof is divided between a bay of regularly-spaced columns and the exposed columns on the exterior, which due to their visual function as supports for the project's graphic identity are not readily identified with their secondary role. This lends a hovering quality to the roof due to the apparent lack of conventional structural support.

The point of entry is marked by a rolling, circular mass that acts as both seal and symbol. Heavy, and sized to cover the circular portal, the door emphasizes the notion of security latent in the project, while reinforcing the underlying geometry of the design. On the inside, a pattern of louvers and circular apertures for lighting accentuate the convex curve of the roof. Extending beyond the point of enclosure, the pattern defines that of the aluminum roof panels, to reinforce the concept of the roof as an autonomous geometric object.

WORLD SAVINGS AND LOAN
ALHAMBRA, CALIFORNIA 2000–2002

世界储备和贷款银行
阿尔罕布拉,加利福尼亚 2000~2002

阿尔罕布拉是洛杉矶市区东部一个中国人聚居的区域，这个银行坐落在该区一条车流量很大的商业街上，以书法为基础的建筑被消减到最基本的形式：一个屋顶，一个外表围合体，一个入口和一个标志。利用铝、玻璃和水泥抹灰为材料，每一个形体都是很基本的几何形，毫不矫揉造作地暗示了建筑的实用性和品牌。

当驱车经过的时候，不同的视角增强了建筑和机动车的动感关系。路旁的标志图案采用建筑外立面同样的材料，与顶部标志图案的两根裸露的钢柱子相呼应。向西倾斜的屋顶当向着车流方向的时候显得像是平的。在视差和速度的限定下，这种三维的建构产生了一系列连续的画面。

支撑屋顶的结构被分成两部分，一部分是由规则开间的柱网，另一部分是室外裸露的柱子。室外柱子因为支撑该建筑的识别图案所造成的视觉功能，而被忽略了它们作为结构支撑构件的角色。这样的非正统结构支撑的做法给屋顶一种漂浮的效果。

一个滚动的圆形的实体标志着入口，它既是门口又是标志。沉重、圆形的大门强调了项目潜在的安全性概念，同时也强化了建筑的几何设计。在内部，屋顶上的百叶组合和圆形的孔洞加强了屋顶上凸曲线的效果。铝板组合的屋顶一直悬挑到墙体之外，加强了屋顶本身是一个独立的几何形体的这一设计概念。

SYLMAR LIBRARY
SYLMAR, CALIFORNIA 2000-2003

萨尔玛图书馆
萨尔玛,加利福尼亚 2000~2003

In an outlying district of metropolitan Los Angeles, set hard against the San Gabriel mountains, the raked lines of the Sylmar Library's metal siding and cellular polycarbonate hug the land like a low-rider. Silhouetted against the peaks, the thin edge of the roof rocks up and down to the same beat as life on the street. It seems like this is a frontier with only a jumble of scrap yards and sun-baked bungalows and helter skelter storefronts, but a hard-working community of mostly immigrant workers fought hard to have a library in the neighborhood.

Hodgetts+Fung's aim was to design a structure that spoke to the land, to the industry of its inhabitants, and to the future of their community. **Hodgetts+Fung** resolved that the character of the building should honor the integrity of manual labor, incorporate references to the local culture, and side-step traditional institutional hierarchies.

structures. Perhaps more importantly, the colors effectively dematerialize the exposed structure, creating an immense surface that has more in common with color field painting than architecture.

The twin circular ducts that crown the entry to the reading room are positioned to act as a portal, yet in their unadorned functionality, they never relinquish their identity as part and parcel of the building's mechanical systems. Similarly, the torqued progression of metal studs supporting the luminous cowl over the circulation counter expresses a raw back-of-house individuality, maintaining a polished yet informal and approachable character.

The building seeks to effect a blurring of boundaries, to yield to impetuous cross-referencing inspired by a mélange of contemporary art, pop-culture, and technology—all tempered by

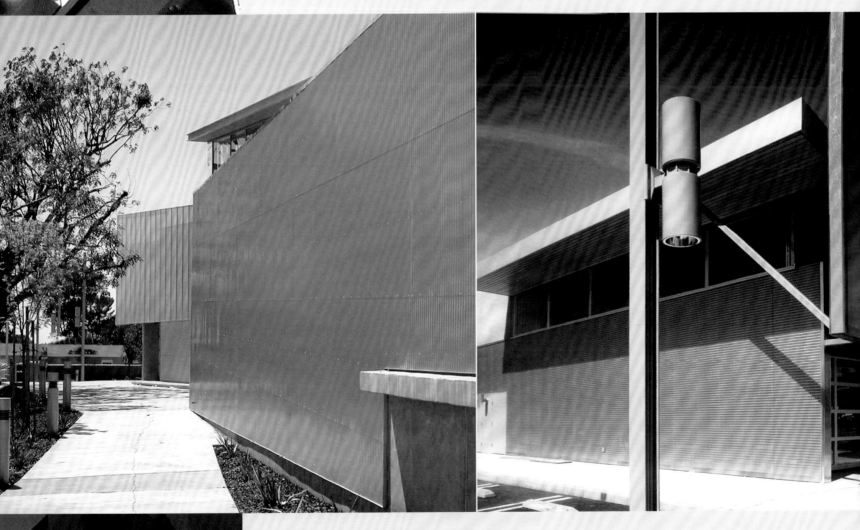

The main space of the library is defined by a casually creased roof resting on irregularly spaced columns, intended to set up a lively rhythm rather than a stately parade. Multi-hued and heavily textured, the roof structure of the library echoes the colors of an eclectic montage of neighboring

simple functional constraints. The goal was a mix of references as casual as the tenor of life on a Sylmar sidewalk, a civic building that is unexpectedly light and unselfconscious, rather than one that demands attention or seeks to dominate the patron's experience.

在大洛杉矶地区的边缘,紧靠着圣加布里尔山,萨尔玛图书馆的倾斜的金属饰板和多孔的聚碳酸酯板俯卧在土地上。薄薄的屋顶边缘的剪影映衬着山峰,上下起伏的频率如同街道上的行人。这个地区看上去像是前线哨所,只有一堆废物场,烈日照晒的单层连廊房子,和乱糟糟的小商店,但是这是一个由多数是外来移民组成的辛勤工作的社区,居民们经过巨大的努力才获得在社区里建一个图书馆的机会。

霍德盖茨+冯的着眼点是设计一个能与环境,居民的产业和社区的未来对话的建筑。他们认为这个建筑的特点应当是推崇体力劳动的荣耀,吸收地方文化,并引用传统的教育结构。

图书馆的主体空间被不规则分布的、由柱网支撑着的一个随意性的皱褶屋顶所围合,这样做的目的是展现活泼的节奏而不是一个呆板的排列。图书馆屋顶结构利用多种颜色和浓重的质感,是从邻近的一个随意性构造物的颜色中得到的灵感。更重要的是,颜色使得暴露的结构材料明确度减弱,创造了一个引人入胜的表面,更像是绘画而非建筑。

一对圆形的管道在入口上方,引导进主要阅览室,如同一个门头。但是它们并没有掩饰它们作为设备系统一部分的这个身分。类似地,支撑圆形服务台上方明亮的灯罩的一组扭转的金属龙骨也表现了一种原始的个性,虽然磨光了,但是保持着随意和近人的特性。

这个建筑设计寻求制造一个边界不很明确的,很强烈地与其他领域相互借鉴的感觉,就好像当代的拼贴艺术,波普文化和现代技术一样,都被简单的功能限定所定义着。因此这个建筑的目标是借鉴一些随意的事物,包括萨尔玛人行道的图案和一个轻巧又不以自我为中心的市政建筑,而不是制造一个寻求人们注意的,强制使用者体验的建筑。

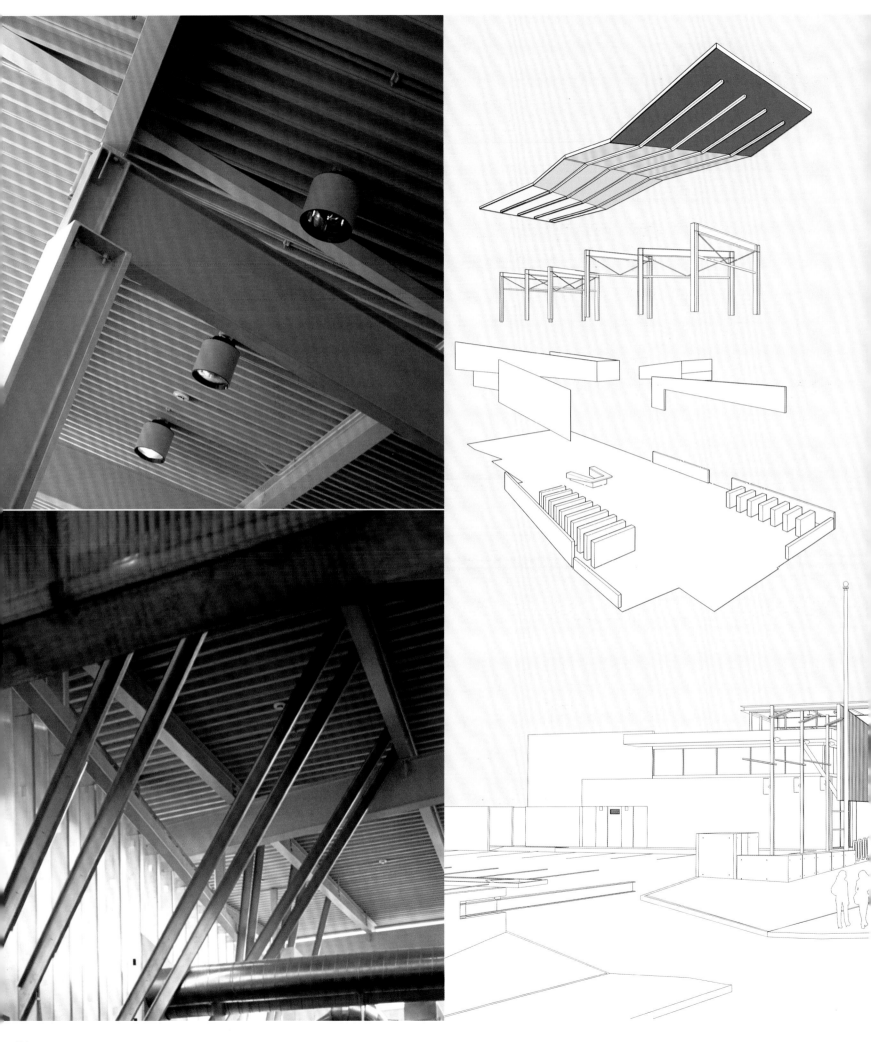

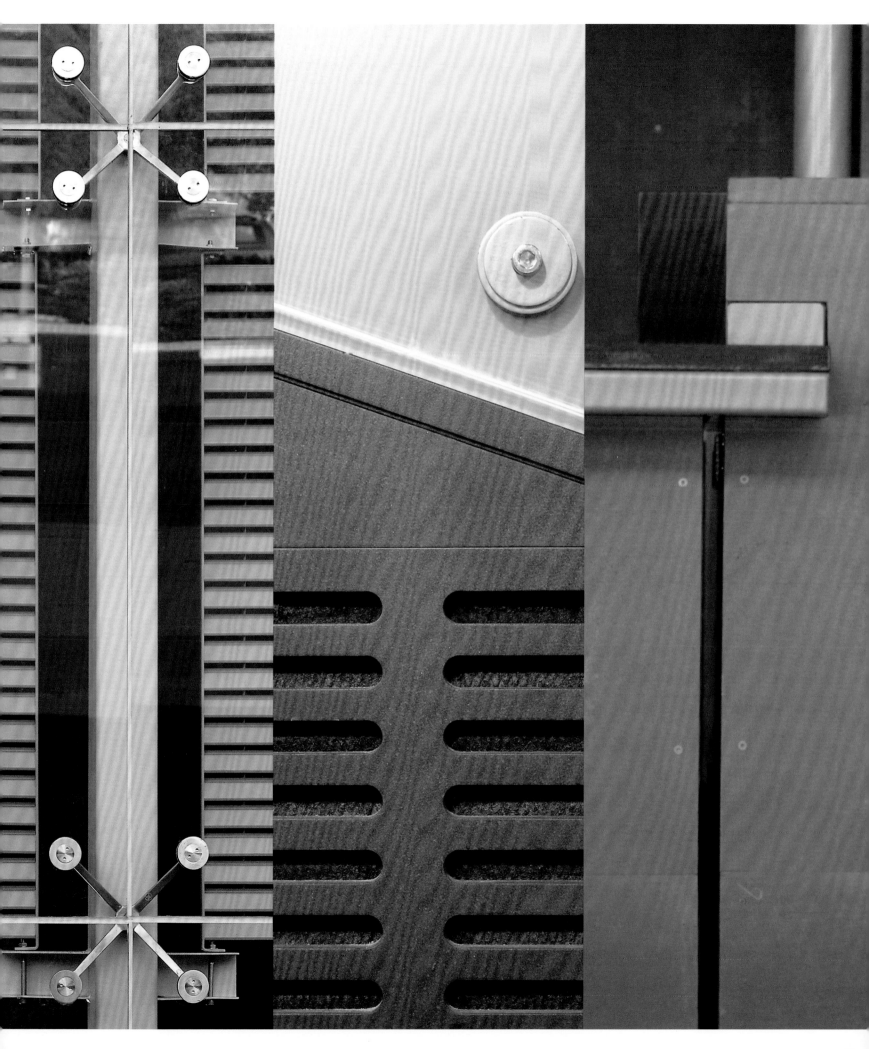

   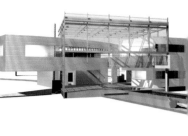

  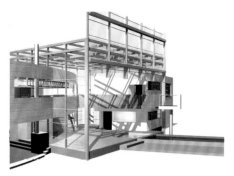 

VIRTUAL HOUSE
SAGAPONAC, LONG ISLAND, NEW YORK 2001–2007

虚拟住宅
SAGAPONAC，长岛，纽约 2001~2007

This summer house was designed to inject the spirit of Southern California lifestyle into the more demanding social and environmental climate of the Sagaponac community. The individual rooms and spaces of the house are arranged informally around a roughly square glass-enclosed courtyard, which forms a cube-like volume designed to capture fragments of the surrounding rooms, spaces, and wooded landscape. The resulting three-dimensional array confers order on an otherwise spontaneous composition, somewhat like the framing device of a photographer's lens.

From the entry, a quasi-circular line of maples describe a long arc that becomes a walkway, then a low wall, then some steps, before sliding into the flank of a glass cube. On re-emerging as a suspended still-curving volume, the arc then descends to embrace space for parking.

The large, family-sized kitchen on the second floor acts as a hub for the activity of the household, linking a home office, the master bedroom suite, a dining balcony, and a panoramic living room to a grand stairway that ascends from the "virtual" space of the ground floor. The flow of indoor to outdoor space is encouraged by the continuity of defining surfaces, which extend the sight-lines into landscaped grounds where echoes of the project's governing geometry may be found.

This is a house that adapts to the seasons. Activated by a unique kinetic element, the cubic living volume acts as a virtual indoor-outdoor porch into which the active elements of the program are projected, enabling the owners to dispense with a traditional enclosure depending on the weather and their whim.

The virtual cube warms under the winter sun, and wakes up to summer breezes when the glass wall to the north is drawn upward to be stored in a special glass housing visible above the roof plane. Balanced by counterweights, and guided by extended tracks, the vertical displacement of the wall redefines the main space to create, in effect, a sheltered, open courtyard bordered by the balconies and entryways, which serve the more traditional rooms of the house.

United by the transparent cube, the individual volumes of the Virtual House bask in light. Each is devoted to a different program—sleep, dining, or play—yet together they are arranged to form a congenial common space. Inward looking, yet expansive in its embrace of the wooded setting, each volume is described by ley lines and arcs delving deep into the forest.

Defined by fragments of surrounding rooms, with glimpses and panoramas visible beyond, the composition rewards a spontaneous lifestyle. Freed from conventional arrangements, a linear pool invites a morning swim, a nearby fireplace inspires conversation, while a library below a sweeping balcony suggests a private read.

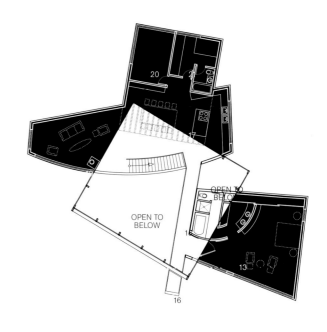

SECOND FLOOR PLAN

13 MASTER BEDROOM
14 MASTER BATH
15 BRIDGE
16 BALCONY
17 KITCHEN
18 FAMILY ROOM
19 FIRE PLACE
20 STUDY
21 MAID'S ROOM

这个夏日度假别墅的设计是把加利福尼亚州南部生活方式的精神注入到Sagaponac社区较为严肃的社交和环境氛围中。住宅的每一个房间和空间都任意地围绕着一个玻璃围合成的近正方形庭院，这个像正方体一样的庭院抓住了周围的房间、空间和树林景观。这样所形成的三维形体产生了一定秩序，否则只是一些随意的堆砌，在某些程度上就像摄影师的取景器一样。

从入口进去，一组枫树成弧形排列，形成一条长长的弧形走道，随后是一个矮墙，几步台阶，然后通过推拉门进入玻璃立方体里。从立方体另一边出来的体量仍然保持弧形，一直下降形成停车库。

二层的一个大家庭式厨房成为家居活动中心，它把办公室、主人套间、一个室外餐厅和一个全景的起居室都联系到一个从底层"虚拟"空间延伸上来的大楼梯。室内外连续性的界面使室内外空间保持流通，并且一直把视线引导到室外景观，建筑的几何形也在那里被重复着。

这是一个适应四季的住宅，立方体的起居室实际上是一个既是室内又是室外的敞廊，通过一个特殊的动能装置和程序，主人可以根据气候的变化把房间包裹上传统的外表。

这个虚拟上的立方体冬暖夏凉，因为北面的玻璃墙可以向上提升，储藏在一个屋顶上的玻璃体里，在热天使空气流通。依靠重锤的平衡和延伸的导轨，这个垂直升降的墙重新定义了主要空间，创造了一个有遮挡的开放庭院，被阳台和门厅等更传统的功能房间相围合。

以这个透明的立方体为中心，住宅不同的房间都沐浴着阳光。每一个房间虽然都服务于不同的功能：

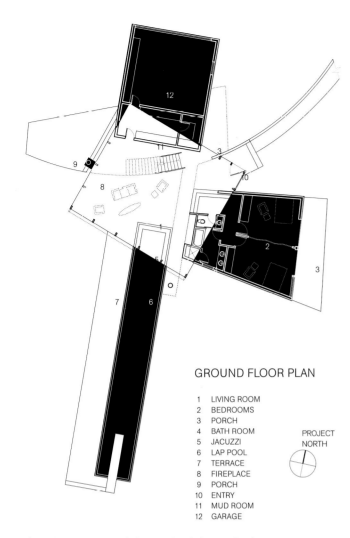

GROUND FLOOR PLAN

1. LIVING ROOM
2. BEDROOMS
3. PORCH
4. BATH ROOM
5. JACUZZI
6. LAP POOL
7. TERRACE
8. FIREPLACE
9. PORCH
10. ENTRY
11. MUD ROOM
12. GARAGE

PROJECT NORTH

睡眠、餐饮或玩耍，但是它们在一起形成了一个天成的公共空间。每个房间都是内向型的布置，同时又延展到周围的树木环境中，建筑体型用直线或弧线定义，延伸到树林深处。

由房间的片断所环绕，间隙中的景观和全景视角所交织，这个住宅的构成反映了一个自发性的生活方式。摆脱了传统的住宅布置的常规，你可以早晨在一个长条形的游泳池里游泳，依靠着旁边一个壁炉闲谈，或是在宽大的阳台下面的书房里独自阅读。

1 COURTYARD
2 STORYTELLING AREA
3 ADULT STACKS
4 TEENS AREA
5 MAIN READING AREA
6 CHILDREN'S AREA
7 NEW BOOKS
8 ENTRY
9 PLAZA
10 MEETING ROOM
11 STAFF AREA

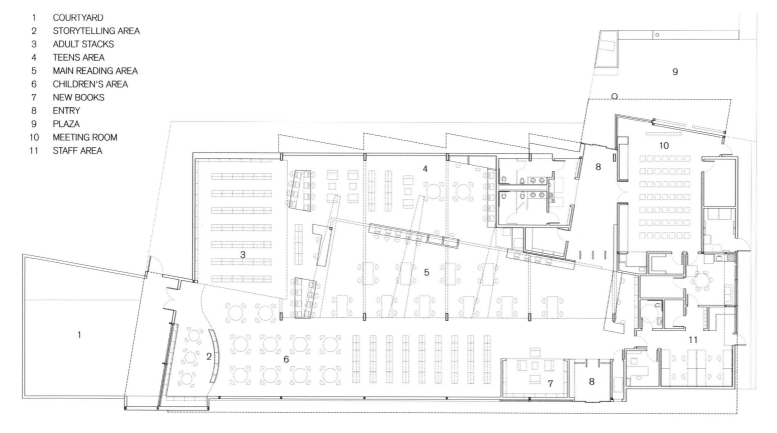

HYDE PARK LIBRARY
LOS ANGELES, CALIFORNIA 2001–2004

海德公园图书馆
洛杉矶，加利福尼亚 2001~2004

Just visible in the midst of the jostling streetscape of its location in South Central Los Angeles, a burnished copper silhouette angles into view. Sun glints from the serrated profile that hovers above the reflection of distant clouds before descending to the sidewalk. Community leaders hope this neighborhood branch of the Los Angeles Public Library system, built on a lot that has stood empty since the violent conflagrations of 1992, will restore pride to the surrounding community.

Hodgetts+Fung sought to develop a syntax of materials and forms founded on the diverse cultural heritage of the community, with the conviction that only by so doing would the library be embraced by its users. The roots of the design lie in the language of African sculpture adopted by modern artists in the 1920s and 1930s, which has been tinged with the bling bling aesthetic identified with hip hop culture. The result is an aesthetic grounded in traditional cultural forms, but enlivened by contemporary experience and tastes, an environment in which the past is at home with the present.

Inside the building, the fabric of the ceiling plane, shot with strands of color and the gleam of irregularly positioned reflectors, turns sharply down from the chiseled profile of roof monitors. Below, pairs of copper-clad ducts project from the walls like spears, stirring the air between them and interacting graphically with the sawn-out profiles of the Paralam bents. Dividing the space and woven from the same synthesis of wood chips and resin as the huge moment frames supporting the roof are computer stations for library patrons.

This is a building of intense contrasts. From the uninterrupted view of the streetscape across Florence Avenue, to the punched-out shards of sky glimpsed through the north-facing lattices, the building makes a darting, almost pugilistic assault on library conventions, yet manages to respond conscientiously to the demands of librarians as well as patrons: unobstructed sight lines, contemplative demeanor, and acoustic ambience are all served. Such a balance between the formal and the informal, the intensely idiosyncratic and the unremarkably mundane, produces a spatial and narrative tension which, admittedly, treads a narrow path, but lends an almost orchestral "color" to the experience of the building.

Along Florence, the library's jig-sawed profile adds a syncopated rhythm to the busy streetscape. Almost crude in its powerful simplicity, yet clad in glistening copper, from afar it looks as if it might be a primitive sculpture made of a precious metal. But where the building meets the ground, parallel to the street and threading behind the copper skin, a moss-green concrete wall encloses an intimate garden for patrons seeking a quiet space for reading.

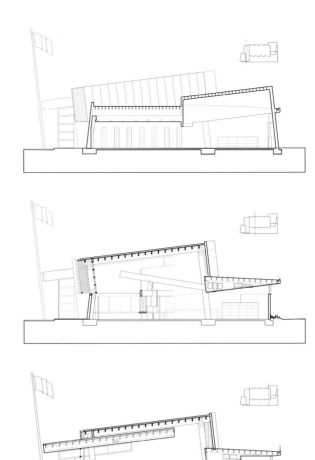
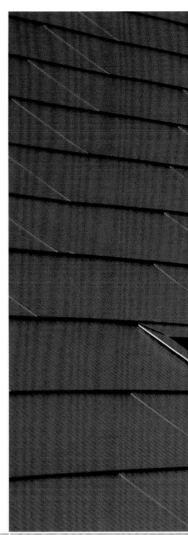

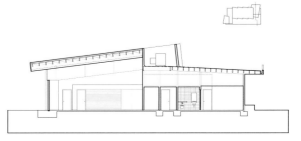

在洛杉矶中南地区拥挤的街道中央可以看见一个折线型的，磨光青铜的剪影，它的锯齿形的轮廓线浮在远处云层的反射光影中，太阳的余晖照在上面闪闪发光。这是一个洛杉矶公立图书馆的分支，建在一片空地上，这片空地自从1992年的暴乱以来就一直是废墟。社区的领导希望这个新的图书馆能恢复附近社区的荣耀。

霍德盖茨＋冯旨在寻找一系列的材料和形式的语汇，与这个社区的多元化的人群文化相协调，因为他们相信只有这样这个图书馆才会被它的使用者所接受。设计的来源是19世纪二三十年代被现代艺术家所采用的非洲雕刻艺术，略带一点Bling Bling 美学和嘻哈文化。设计的成果是美学植根于传统的文化形式，但是又有着当代的体验和格调，是一个既有故乡情结又有现代感的环境。

在房子里面，顶棚平面的织物掺杂着彩色条纹并且不规则地镶嵌着闪烁的反射镜，从轮廓鲜明的屋顶太阳能装置悬挂下来。在下面，一对对的铜质通风管像矛一样从墙上伸出来，用以调节空气，并在形式上和它的特殊转接弯头相呼应。图书馆用户的电脑设施被用合成的木屑板和树脂围合起来，同样的材料也用在铰接屋架上。

这是一个含有极强烈对比的建筑。从沿着佛罗伦萨大道连续的街景到北面方格围栏孔隙中透过的蓝天，这个建筑对传统的图书馆建筑的常规的打击像拳击一样猛烈，然而在同时又能满足图书管理员和使用者的要求：无遮挡视线，激发思考的气氛，声学漫射等都予以实现。这样一个正规和随意、强烈的个性化和极度平凡之间的平衡，产生了一种空间和表述性的张力，它经过一个狭窄的小道但是又带来一种几乎是管弦乐般色彩的对建筑的体验。

在佛罗伦萨大道上，图书馆锯齿般的形状给忙碌的街道增添了一种变奏的韵律。几乎是残酷的简单，然而包着发光的青铜，从远处看好像一个有稀有金属做成的雕塑；但是在建筑接触地面的底层，与街道平行并设在铜外皮的后面，一片苔绿色的混凝土墙围合了一个花园，使用者可以在其中找到一块安静的地方阅读和思考。

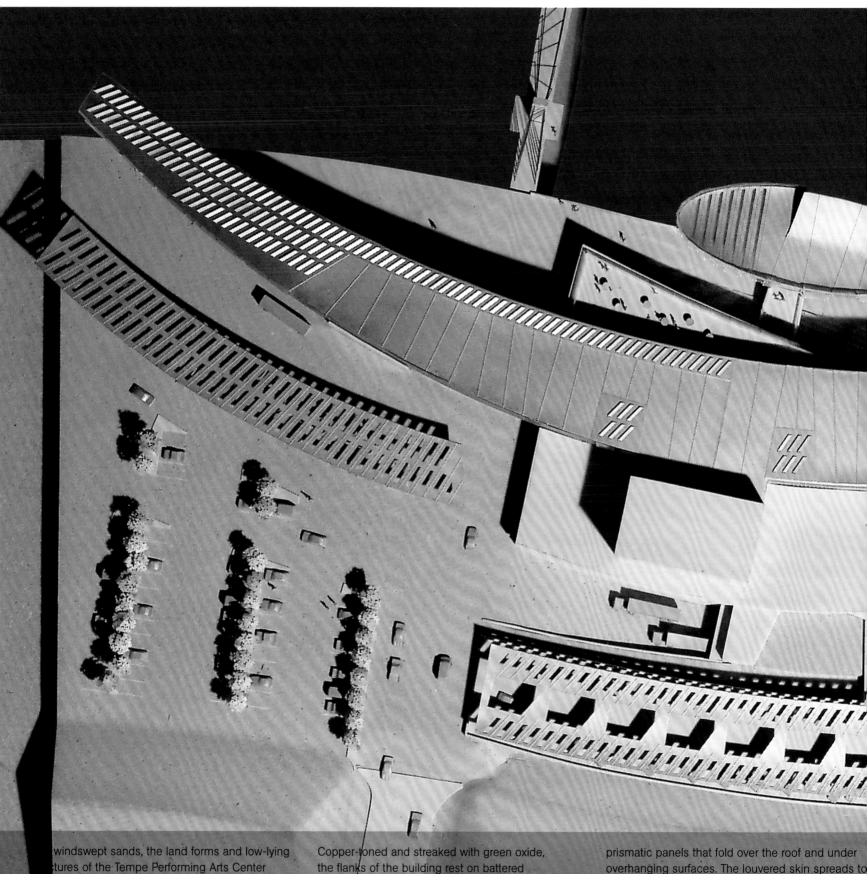

...windswept sands, the land forms and low-lying ...tures of the Tempe Performing Arts Center ... the desert landscape where it meets the ...ntly flooded Rio Salado. Organized along a ...ng spine, theaters, a gallery, a cinema, and art s...os share a two-story lobby overlooking a new m...-made lake as well as an elevated outdoor te...ce oriented toward the setting sun.

Copper-toned and streaked with green oxide, the flanks of the building rest on battered walls composed of local sandstone. A glowing tower marks the stage, warding off the sound of ascending aircraft with a double, internally-illuminated skin. The skin of the structure is "gilled" with louvers made of corrugated copper cladding, which in turn is laid in irregular, quadrilateral

prismatic panels that fold over the roof and under overhanging surfaces. The louvered skin spreads to form a monumental trellis, creating a grand, shady outdoor "room" that functions as a gathering place for school children and tourists. A relaxed curve defines the edge between building and sky, and draws a sharp contrast with the jagged formations of desert rock across the lake. Busses and cars car...

TEMPE VISUAL AND PERFORMING ARTS CENTER
TEMPE, ARIZONA 2002

坦佩视觉与表演艺术中心
坦佩，亚利桑那州 2002

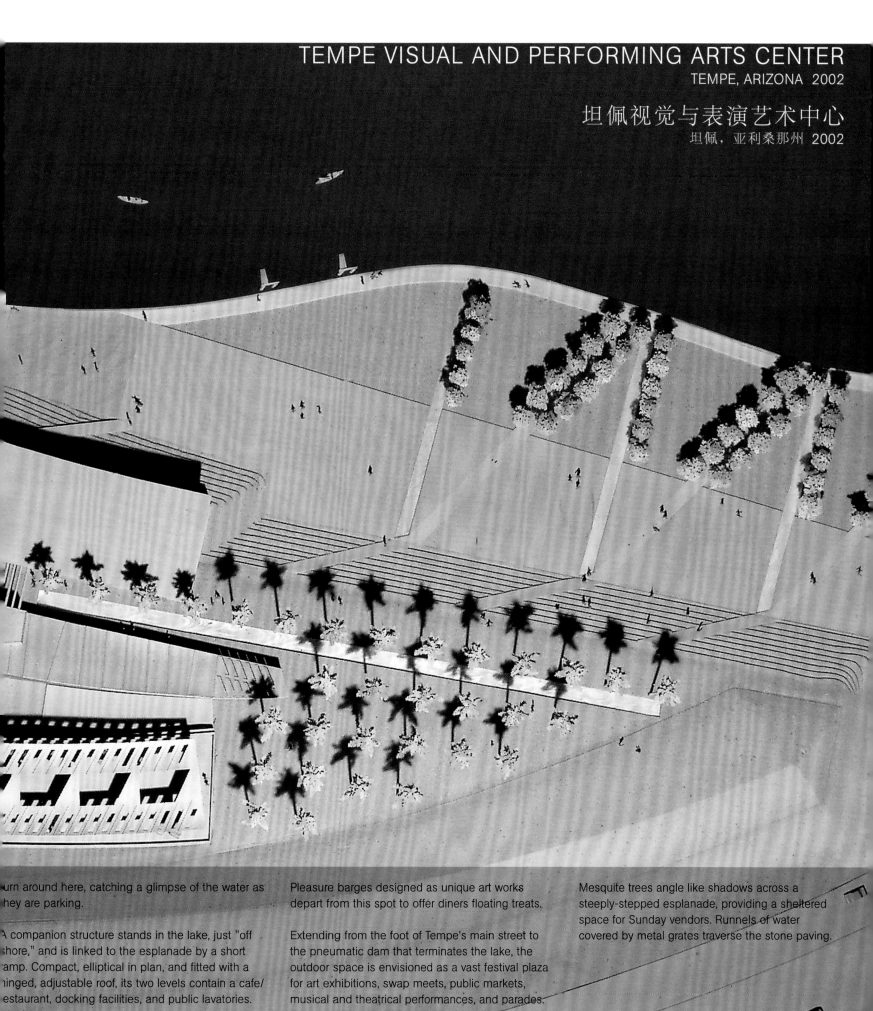

urn around here, catching a glimpse of the water as hey are parking.

A companion structure stands in the lake, just "off shore," and is linked to the esplanade by a short amp. Compact, elliptical in plan, and fitted with a inged, adjustable roof, its two levels contain a cafe/ estaurant, docking facilities, and public lavatories.

Pleasure barges designed as unique art works depart from this spot to offer diners floating treats.

Extending from the foot of Tempe's main street to the pneumatic dam that terminates the lake, the outdoor space is envisioned as a vast festival plaza for art exhibitions, swap meets, public markets, musical and theatrical performances, and parades.

Mesquite trees angle like shadows across a steeply-stepped esplanade, providing a sheltered space for Sunday vendors. Runnels of water covered by metal grates traverse the stone paving.

WATER

STRUCTURE

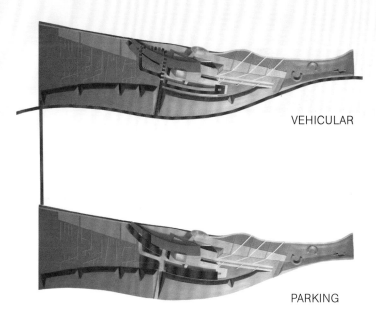
VEHICULAR

PARKING

像风吹过的流沙，在沙漠和最近发洪水的里约萨拉多河交界的地方，低矮的坦佩视觉表演艺术中心建筑和它周围的地貌共同形成了沙漠景观的边界。沿着一根弯曲的脊柱分布的剧场、画廊、电影院和艺术工作室共有一个临水的两层高门厅，门厅俯瞰着一个新造的人工湖和一个面向落日的，抬高的室外平台。

建筑的侧面建在用当地的砂岩砌成的倾斜的墙上，青铜色调略带一点绿色氧化物。作为标志物的是一个发光的塔，用一种双层的，内部照明的外表材料来避开飞机起飞的噪声。建筑的外表覆盖着波浪形青铜覆面的百叶，这种青铜百叶也用在屋顶上，排列成不规则的、四棱柱形的面板；同时也用在悬挑屋檐的底面。再进一步，青铜百叶被用于一个有纪念性的格架，形成了学生和游客聚集用的一个庞大的、阴凉的室外"房间"。一个很舒缓的线条分界着建筑和天空，这与湖对面沙漠中凹凸不平的岩石形成鲜明的对比。公共汽车和小汽车可以在这里转弯，泊车的时候还可以看见湖面。

在湖中有一个附属构造物，正好在水边，由一个短坡道与主要广场相连接。小巧的，椭圆形的平面，顶部一个带转轴可以调节的屋顶。这个两层高的小品包括一个咖啡/餐厅，停船设施和公共卫生间。导览的驳船设计成一个独特的艺术品从这里起航，游客可以在船上用餐。

室外环境从坦普的主街一直延伸到拦截湖水的气压水闸，可以作为艺术展览、交易会、公众市场、音乐戏剧表演和游行使用的巨型广场。牧豆树成角度排列，如同影子一样穿过一个坡度很陡的台级式的广场，给星期天来这里的小商贩提供了遮荫的地方。盖着金属箅子的明沟横穿过石材铺地。

LANDFORMS

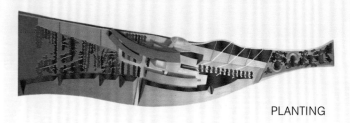

PLANTING

PLAZA

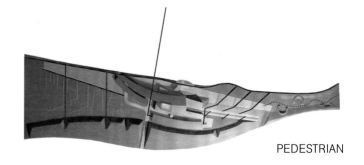

PEDESTRIAN

8:00	8:15	8:15	8:20

Staff and visitors arriving at the parking area to the north of the building cross to a tree-lined walkway to begin their day at the Endowment.	Approaching the entrance, they might have time to join co-worker a they see sipping an espresso on the terrace above.	Those having space assigned in the underground garage can take an elevator directly to their floor, or enter the courtyard by ramp or stair.	Leaving the underground parking area, Endowment staff pass through a garden terrace and pool before arriving at the courtyard level.
员工和参访者到达建筑北侧的停车场，然后穿过一个林荫道去开始他们在捐赠协会一天的生活	到达入口后，也许他们还有时间和同事在上面的平台上享用一杯浓咖啡	人们可以通过地下车库的电梯到达他们所在的楼层，也可以进入天井从楼梯上去	离开地下停车场，捐赠协会的员工在到达天井前，他们会路过一个花园露台和游泳池
12:30	12:30	16:00	18:00

Those who take lunch on the cafe terrace are sheltered from noise by the sound of water rippling over the fins of a specially-designed chiller.	while those on their way to Chinatown or Olvera Street pass through a busy public corner with a flowing fountain and broad benches.	For an event at the close of the business day, the skyline of Los Angeles helps to animate the gathering, while a relaxed trellis provides shade.	A 180-degree view from the boardroom embraces the diverse communities of the city to form a panoramic backdrop.
员工在咖啡露台上享用午餐时，所有的噪音都被特别设计的响水幕帘给屏蔽掉了	人们在步行去中国城和奥尼尔大街时，将穿越一个带有喷泉和长凳的繁忙公共角	在一天的工作时间快束时去参加一个活动，这时洛衫矶的天色松散的照射进来，给聚会创造了良好的气氛	180度视野的会议室可接纳不同的公众人群，这也是一个城市的一个缩影

A competition for the headquarters of this private, statewide multicultural medical foundation prompted a design inquiry into the nature of the contemporary corporate "campus." Beginning with the conviction that an enlightened corporation should provide a working environment that nurtures the creativity and individuality of its employees, and recognizing that an increasing reliance on computing and information technology has eroded the quality of face-to-face encounters, it became clear that the modernist office design template had become dysfunctional—a breeding place for bureaucracy instead of a breeding place for ideas.

Chief among the client's goals for the project was the enhancement of productivity by the provision of out-of-the-ordinary and memorable places for routine meetings, places in which location, configuration, and even material vocabulary would surprise and stimulate interaction. It was hoped that a strategy of intentional disparity between building components would reinforce employees' sense of individuality, and thus empower the California Endowment with a revitalized staff.

Adapting the principles of courtyard housing typical of Los Angeles to this larger and more complex mission, **Hodgetts+Fung** envisioned a synergistic workplace charged with the energy of spontaneous encounters, where creativity, teamwork, and loyalty replace traditional corporate competitiveness. Their approach assumed that, while communication, data transfer, and management structures will continue to evolve in an unforeseeable manner, the fundamentals of face-to-face meetings and real-time discussion will continue to thrive in spaces that lend meaning and substance to the matters at hand.

All the activities that take place in the campus setting are grouped around a courtyard, which is modulated in both the horizontal and vertical dimensions by generous stairs and surrounding terraces. Meeting rooms, the library, a staff lounge, and the cafeteria have immediate access to these common areas, providing multiple opportunities for break-out sessions, and enabling participants to retreat to a balcony, or conclude their discussions in a flower-filled greenhouse.

THE CALIFORNIA ENDOWMENT
LOS ANGELES, CALIFORNIA 2002
加利福尼亚捐赠协会
洛杉矶 加利福尼亚州 2002

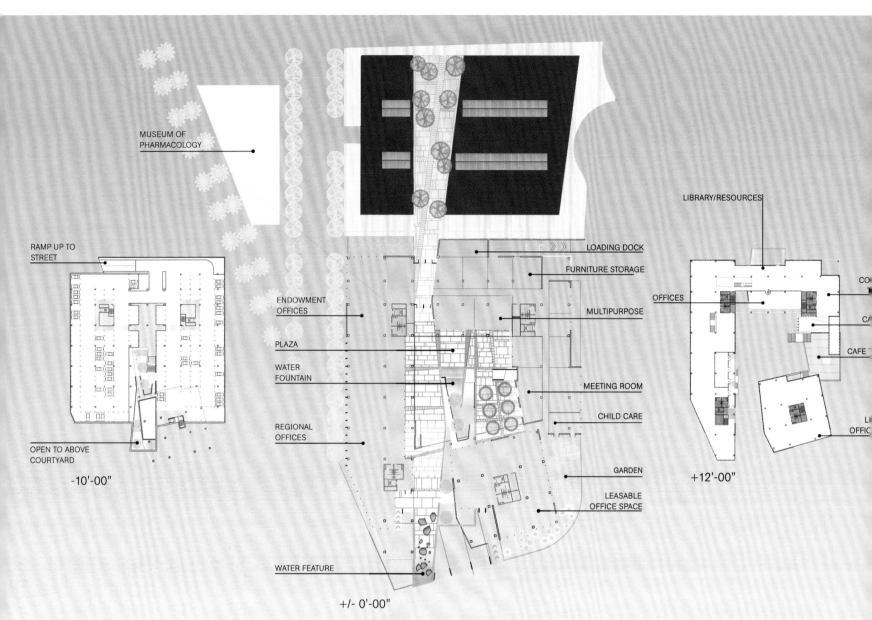

这是一个私人拥有的州立多文化医疗基金会。它的总部的竞赛引发了一个当代公司集团"校园"的设计问题。首要的观点是具有启迪性的集团公司应当提供一个培养职工创造力和个性化的环境，然后是意识到不断增长的对计算机和信息技术的依赖性减损了面对面交流的机会，一个很明确的事实是现代化的办公设计模式已经造成了功能的误区——一个制造官僚主义的地方而不是一个产生思想的地方。

业主对项目最重要的目的是通过一系列的规定来提高工作效率，包括不寻常的有意义的会议场所，其他在地点、组织甚至材料应用方面都有新奇感，可以激发互动的场所。希冀一个有意区别建筑元素不同性的策略可以加强职工的个性感，由此可以强化加利福尼亚捐赠会人员的活力。

霍德盖茨＋冯采用洛杉矶地区庭院式住宅的原理来设计这个大得多的、更复杂的综合体，希冀创造一个通力合作的工作空间，它充满了自发性的交往活力、创造力、集体合作和企业忠实性，而代替了传统的企业竞争性。这样的方式可以表述为，即使信息交流，数据库传送和管理结构仍会继续以一种看不见的形式参与，面对面的交流和实在的讨论这个基本原则将会在那些有意义和实质性设计的空间中得到发扬光大。

所有规划中的行为都沿着一个庭院组织布置，这个庭院用宽阔的楼梯和周围的平台在水平上和竖直上统一起来。会议室、图书馆、一个员工沙龙和咖啡厅都直接和这些公共空间相连接，提供了更多的休整的机会，使得参与者可以在阳台上或在充满鲜花的花房里结束他们的讨论。

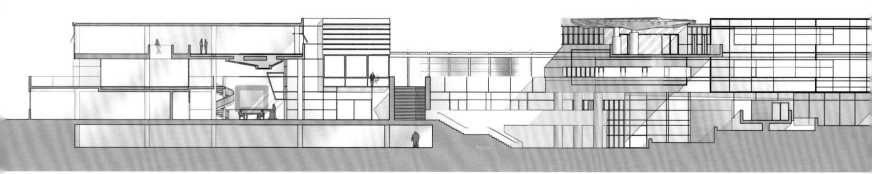

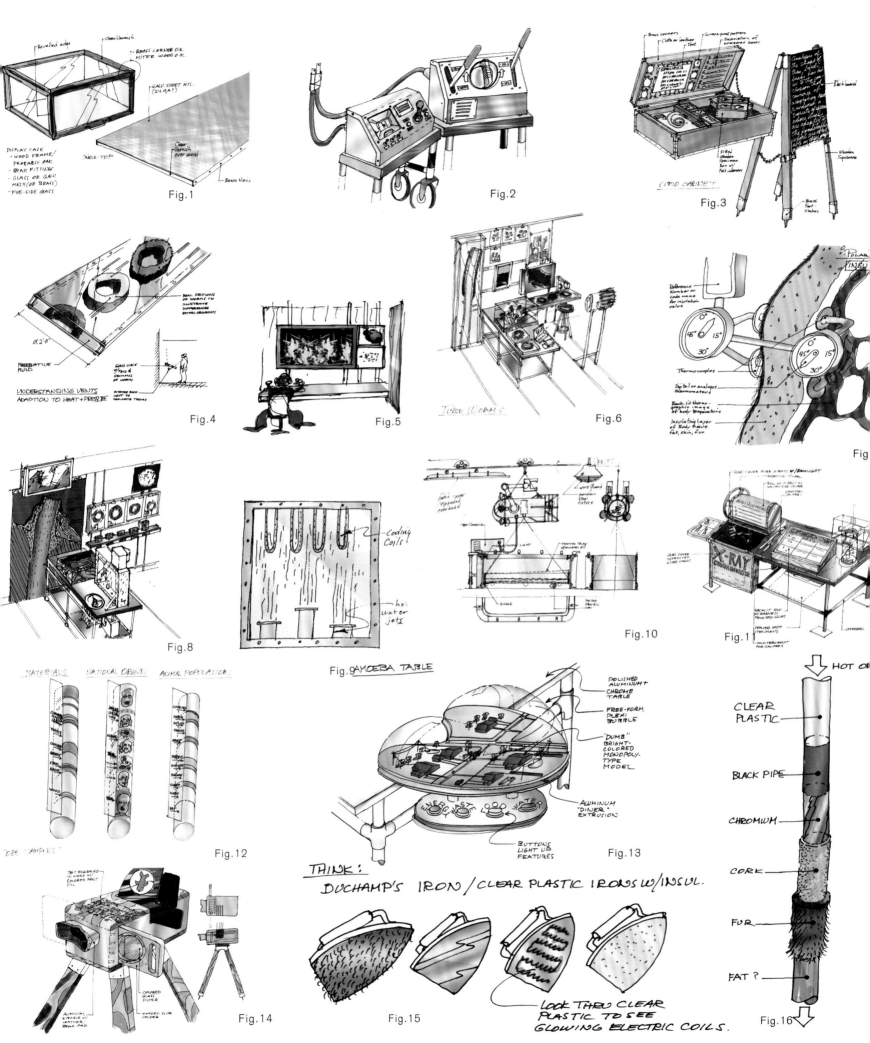

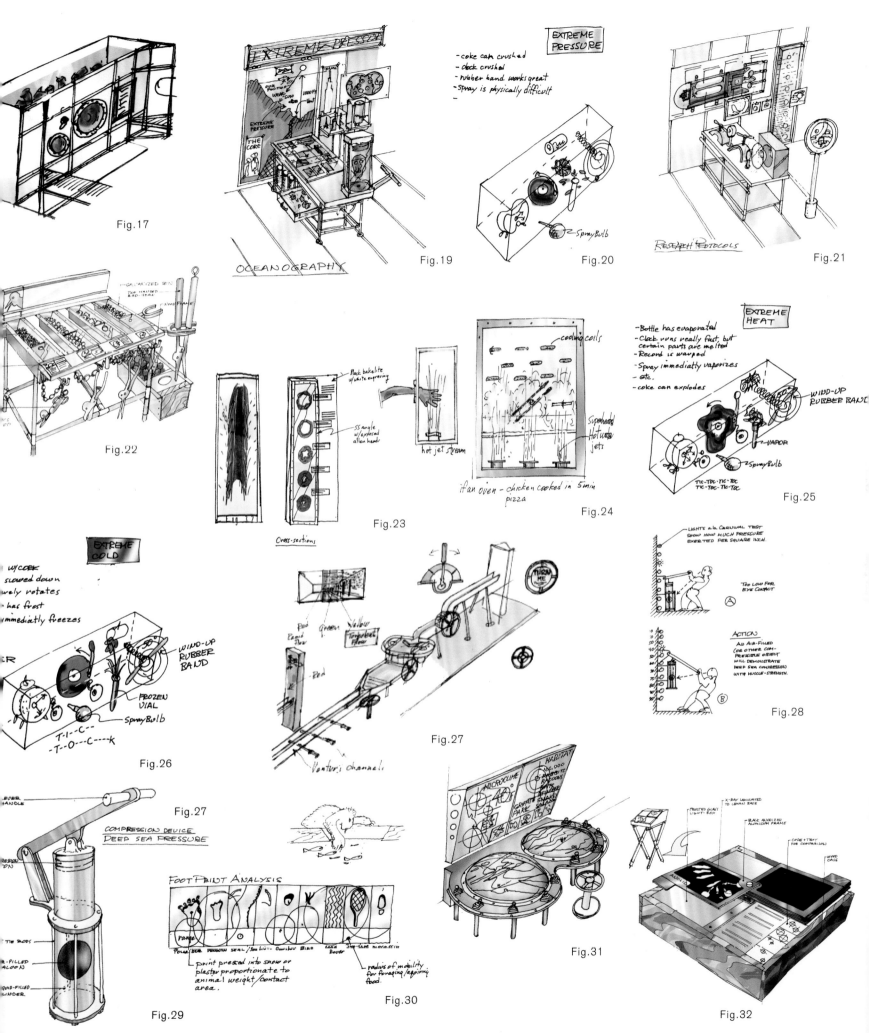

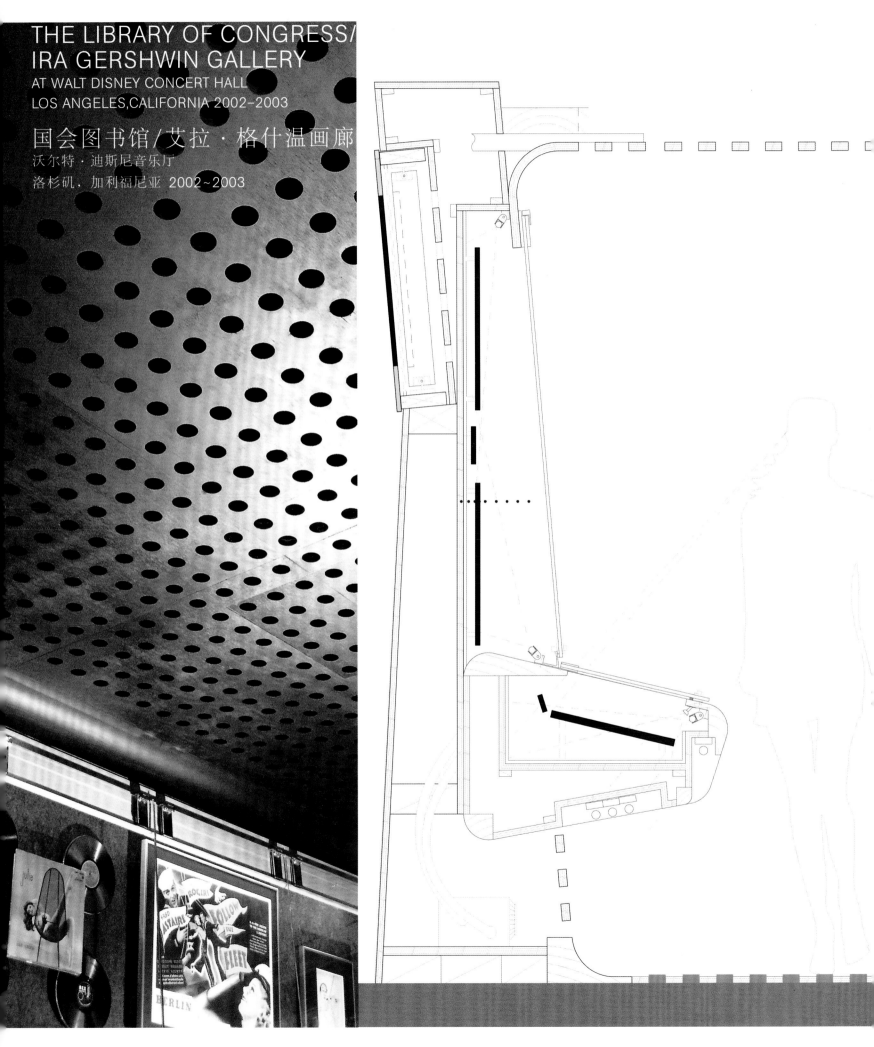

THE LIBRARY OF CONGRESS/
IRA GERSHWIN GALLERY
AT WALT DISNEY CONCERT HALL
LOS ANGELES, CALIFORNIA 2002–2003

国会图书馆/艾拉·格什温画廊
沃尔特·迪斯尼音乐厅
洛杉矶，加利福尼亚 2002~2003

Conceived as an exhibition space for concert-goers and daily visitors to appreciate musical scores, photographs, and ephemera of the Los Angeles Philharmonic Orchestra, the Gershwin Gallery adapts the materials and motifs of Frank Gehry's Walt Disney Concert Hall to a more intimate key. The first impression is one of differences in scale. Here, in a tiny space compressed between two larger areas, the ubiquitous perforated plywood, stainless steel, and multi-colored carpet of Disney Hall are suddenly amplified, like a jump-cut from a distant landscape to a close-up of a single flower within that landscape.

Carved into the void between the undulating walls of the Disney Hall lobby, the diminutive Gershwin Gallery is sandwiched between the grand concourse and the recital stage, offering an ideal opportunity for a moment's diversion pre- or mid-concert. To protect the artifacts culled from the music and historical collections of the Library of Congress, the gallery is built to museum specifications, with low-level fiber-optic lighting and sophisticated temperature and humidity controls.

The resulting sharp contrast between the illumination of the public areas of the Concert Hall and that of the Gershwin Gallery is mediated by a two-tiered approach to lighting. The fact that the gallery is approached up a slight incline made it possible to create an intriguing patterned glow that is only visible from below, and yet maintain the low light levels required within the display cases.

A faceted stainless steel entry wall burnished to echo the exterior of Gehry's building is configured as a prismatic solid to accentuate the flamboyant curves of the concert hall. Large drilled holes mimic perforations of the lobby walls. This motif is continued in the ceiling, which provides acoustic damping, and repeated in the flooring, where black neoprene roundels make a cushioned surface. Massive rectangular Douglas Fir display cases contrast with the curving panels surrounding the gallery. Objects float on a dark felt-grey surface beneath glass that has been sloped to avoid reflection. This is a nearly barrier-free environment for the appreciation of artifacts.

格什温画廊是一个给去听音乐会的人和游客提供的展览空间，他们可以在其中欣赏乐谱、照片和洛杉矶爱乐乐团的海报。它的设计把弗兰克·盖里的音乐厅的材料和母题用于更亲近的感觉。首先的印象是尺度的不同，在这里，两个大空间之间的小空间里，迪斯尼音乐厅那无处不在的打孔三合板、不锈钢和五彩地毯突然被放大了，就像从一个远景跳到风景中的一朵花的近景一样。

格什温画廊是从音乐厅门厅波浪形的墙体之间空的部分切割出来的，不起眼地夹在大厅和排练舞台之间，在演出前和演出休息中是一个理想的去处。为了保护从国会图书馆音乐和历史收藏部挑选出来的艺术品，这个画廊以博物馆标准设计，采用低度的光电采光和复杂的温湿度调节。

音乐厅和格什温画廊照明之间产生的巨大差异也被两层灯光所调节。画廊事实上是在一个稍微高的位置上，这样一来灯光的设计可以较为独特，只能从下面看到，所以可以再按要求保持低度的照明。

画廊入口处的墙采用磨光的不锈钢饰面，回应着盖里的外立面特点。它被组合成三棱柱实体的形状以加强音乐厅炫耀的曲线。一系列的大钻孔是模仿音乐厅门厅的打孔板。这个母题继续用在顶棚上来减弱噪声，用在黑色橡胶地面的圆形突起，有减震垫的作用。大面积的枞木演示台和周围弧形的墙面形成对比。展示物品摆在玻璃下面的黑灰色的台面上，台面略微倾斜以防止反射光。这个画廊设计是无障碍的。

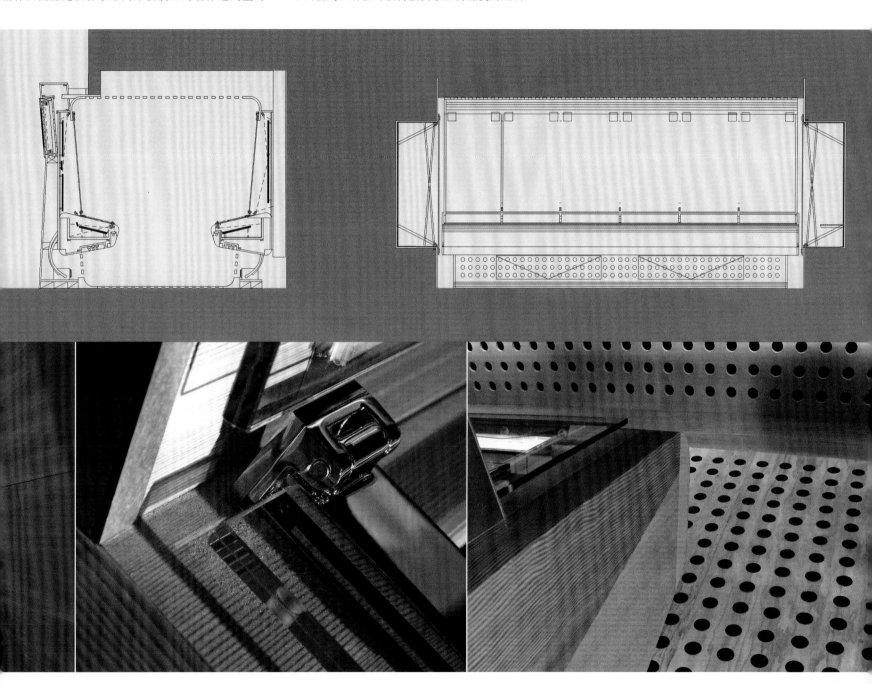

PARQUE DE LA GAVIA
MADRID, INVITED COMPETITION 2003

加维亚公园
西班牙马德里，设计竞赛邀请 2003

Drawing on memories, unfulfilled dreams, and films in their project for an invited competition to design this park site south of Madrid, **Hodgetts+Fung** used mounds of reclaimed soil as the basis for creating six distinct zones of striated landscaping: forest, shadowed berms, terraced gardens, urban plaza, sculpted lawns, and wild grasses. They proposed scenarios as diverse as Thoreau's "Walden Pond" and Antonioni's "Red Desert" in order to tempt the typical visitor to seek out and find the "zone" that matches their mood or temperament: solitude, searching, engagement, socializing, strolling or playing. The park environments envisioned in this scheme exist as independent phenomena tailored to support a specific set of expectations. They reward the visitor with unique and provocative combinations of trees, flowers, shrubs, landforms and patterns, pavings, and other furnishings.

Density, porosity, order, and technology are admixed in varying proportion, guiding the purification of ground water, the propagation of plants and wildlife, and the enjoyment of nature modified by artificial and even theatrical devices. The vessels that conduct water through the process of sedimentation and oxidation are arranged to underscore the complementary roles of nature and technology. Thus the water, which is seen at first descending through huge antiseptic stainless steel pipes connected to an enormous stainless steel tub, is later seen flowing through concrete sluices, until it eventually cascades over broken stone ledges.

The project for Parque de la Gavia grapples with issues of water reclamation and management, from the natural origins of streams and rivers to grand, man-made urban waterscapes. It engages environmental responsibility, education, and the recreational use of nature in varying proportions, advancing a polemic agenda wherever possible. Such a re-imagining of the landscape, the treatment of all things as part of an expanded ecology, is simply a reflection of our contemporary condition, in which the blurring of boundaries between the natural and the synthetic offers rewards to the environment as well as the public. Walls of recycled and compacted aluminum, plastic, and steel waste hold back flowerbeds and sediment ponds. The colorful motion of wind turbines powers the pumps that recycle storm water. Cast-off automobile windshields dot the landscape and provide shelter from wind and rain.

If we are to survive our own success, we must turn our attention to the integration of the natural world with our technological powers and stop thinking of them opposed to one another. Above all, every manipulation and transformation of the landscape must re-imagine it in terms of interlocking cultural, natural, and technological systems that address the landscape not as a static, passive ground but as a dynamic space in the process of transformation.

霍德盖茨＋冯以记忆、未实现的梦想和电影为素材创造了这个马德里南部的公园设计邀请竞赛。他们用开垦的土堆作为基础设计了六个截然不同的条形景观区：森林、树荫下的山坡、平台花园、城市广场、雕塑草坪和野草。他们畅想的情景也截然不同，就好像梭罗的"瓦尔登湖"和安东尼奥尼的"红色地毯"一样的差别，目的是吸引一般游客寻找与他们的情绪或气质相似的区域：孤独、寻求、约定、社交、漫游或是玩耍。在这个设计里每一个公园的环境被视为独立的现象存在，经过裁减后维持着一系列特别的想像。他们同时提供给游客独特并且刺激想像的组合元素：树、花、灌木、地貌和图案、铺地和其他装饰。

密度、多孔性、秩序和技术用不同的比例混合，指导地下水的纯化、植物和野生动物的繁殖，以及对人工甚至戏剧化地改造的自然的欣赏。引导水流的装置经过沉淀和氧化过程被刻意布置以强调自然和技术的补充关系。水流起初向下通过粗大的防腐蚀不锈钢管引导到一个巨大的不锈钢水池中，后来可以看到水流经混凝土水闸，直到最终瀑布从裂开的石槽落下。

加维亚公园方案设计在水的利用和管理问题上颇费脑筋，从自然发源的小溪、河流到人工的城市水景设计。它以不同的比例结合了环境保护义务，教育和自然的重复利用，尽可能地使这些带争议性的议程又更进一步。这样一个重塑的景观，把所有问题的解决方法都作为可拓展的生态学一部分，是我们当代境况的一个简单的反映，即自然和人工之间的界限渐渐模糊了，它对环境和公众的积极作用是一样的。用回收的铝制品、塑料制品和钢所制成的墙可以用做花坛和沉积池；绚丽多彩的风向标可以带动马达来回收雨水；废弃的汽车挡风玻璃可以装点景观用来避风遮雨。

如果我们能从我们自己的成功中存活出来，我们必须把注意力放在如何使自然和人类的技术力量协周起来，而不应当想像它们是对立的。在所有基础上，对景观的每一步操纵和整形时都应当考虑相关的文化、自然和技术系统，它们能否使景观变成一个有活力的场所，而不是一个静止的、消极的地方。

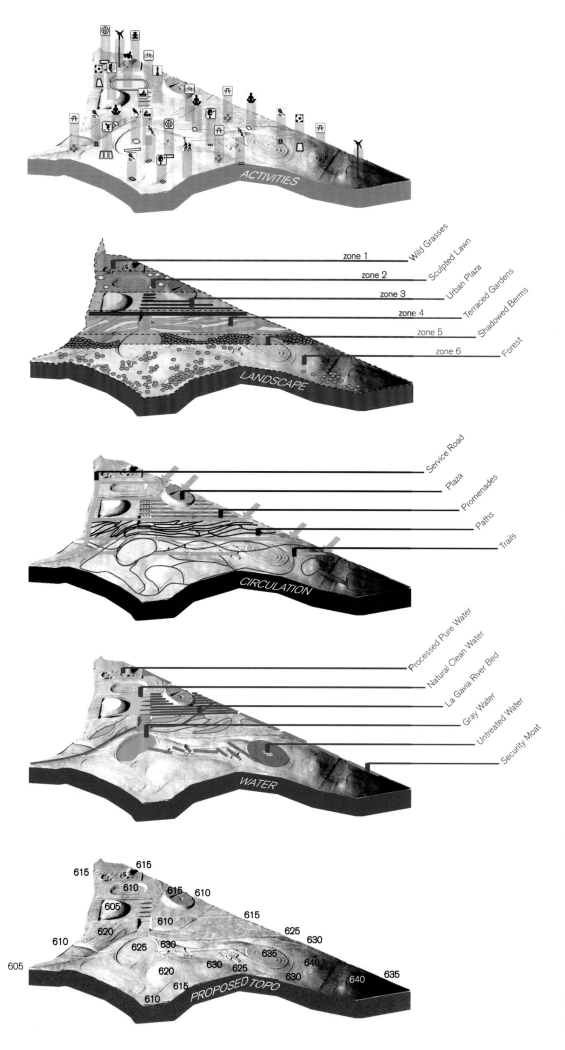

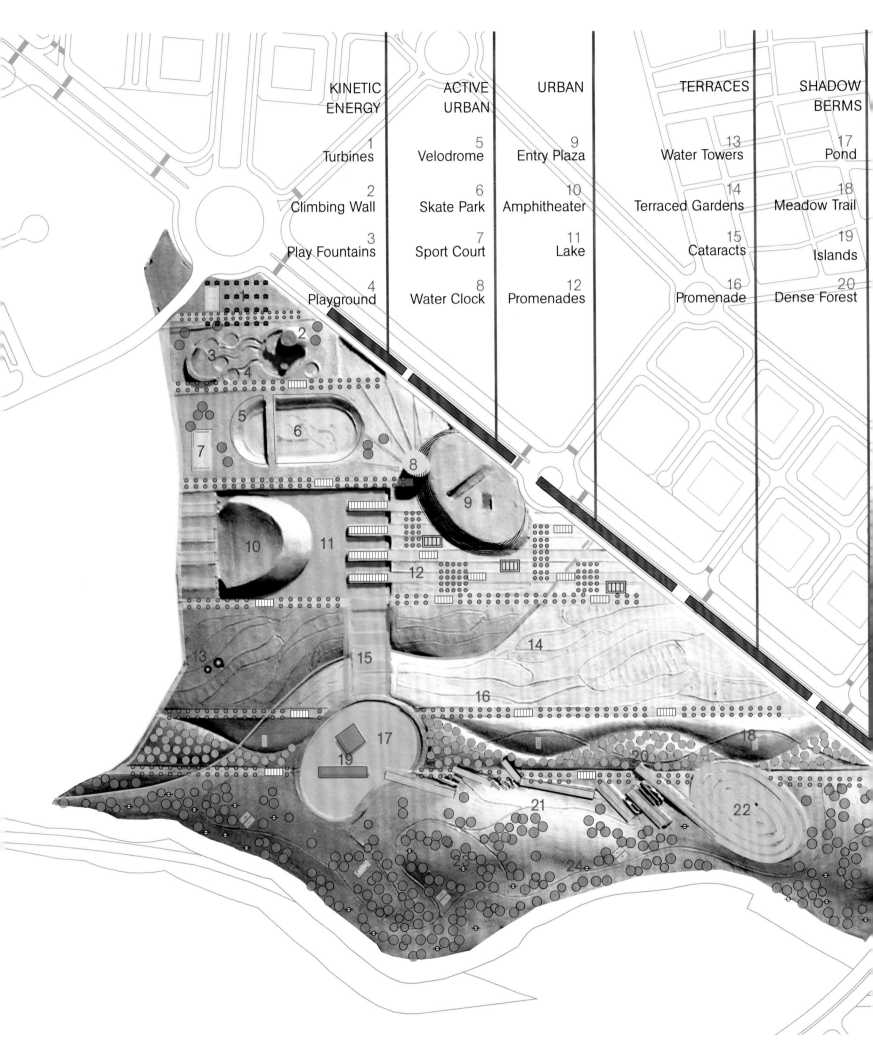

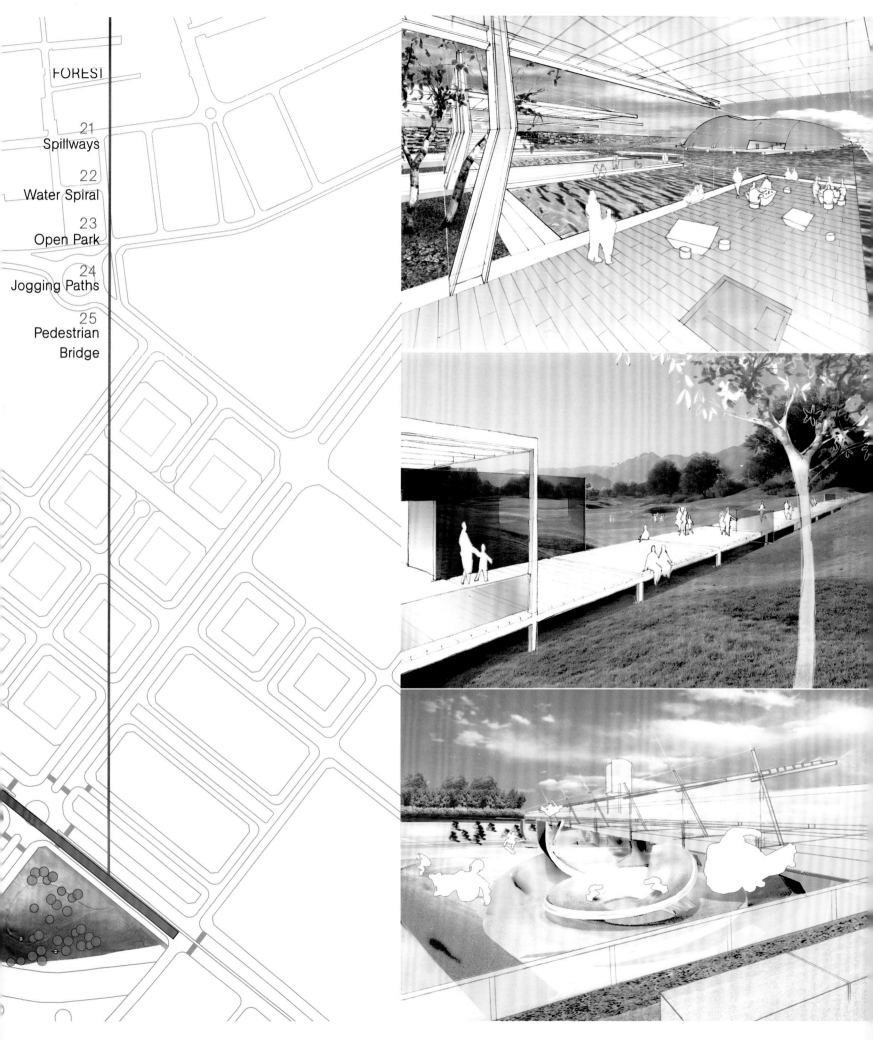

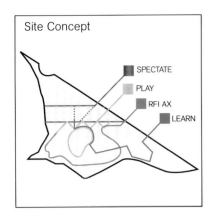

Site Concept
- SPECTATE
- PLAY
- RELAX
- LEARN

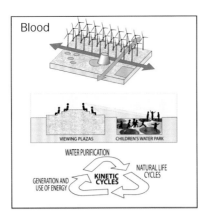

Blood

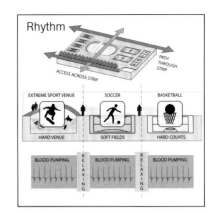

Rhythm

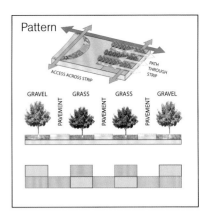

Pattern

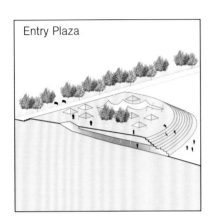

Entry Plaza

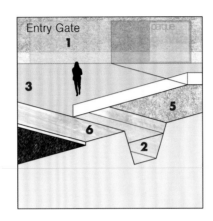

Entry Gate

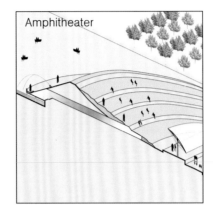

Amphitheater

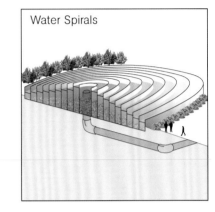

Water Spirals

Recycled Copper Wires

Recycled Cars

Recycled Cans

Recycled Tires

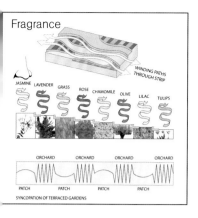
Fragrance

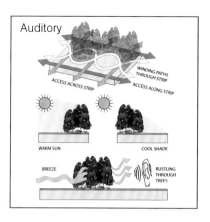
Auditory

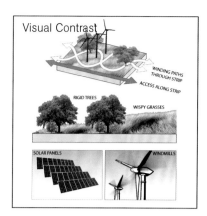
Visual Contrast

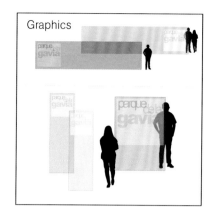
Graphics

Trellises

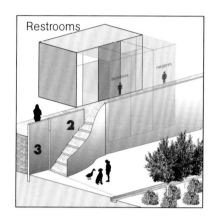
Restrooms

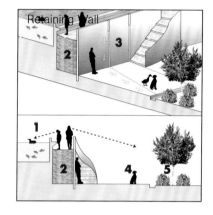
Retaining Wall

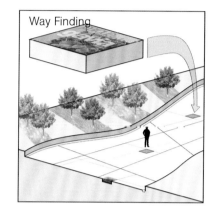
Way Finding

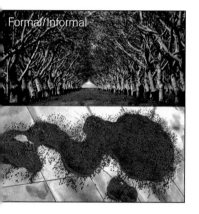
Formal/Informal

Colorful/Monochromatic

Shadow/Sun

Static/Movement

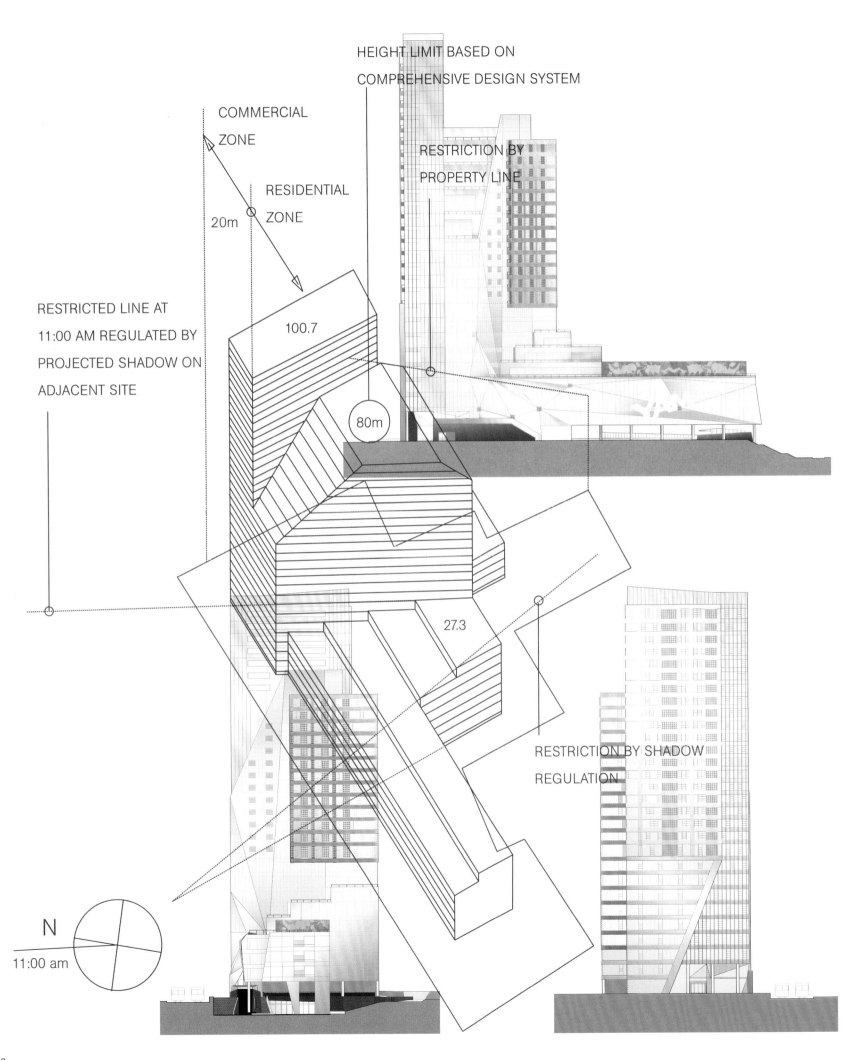

YAMANO GAKUEN
TOKYO, JAPAN 2003-2007

东京山野学校
东京，日本 2003~2007

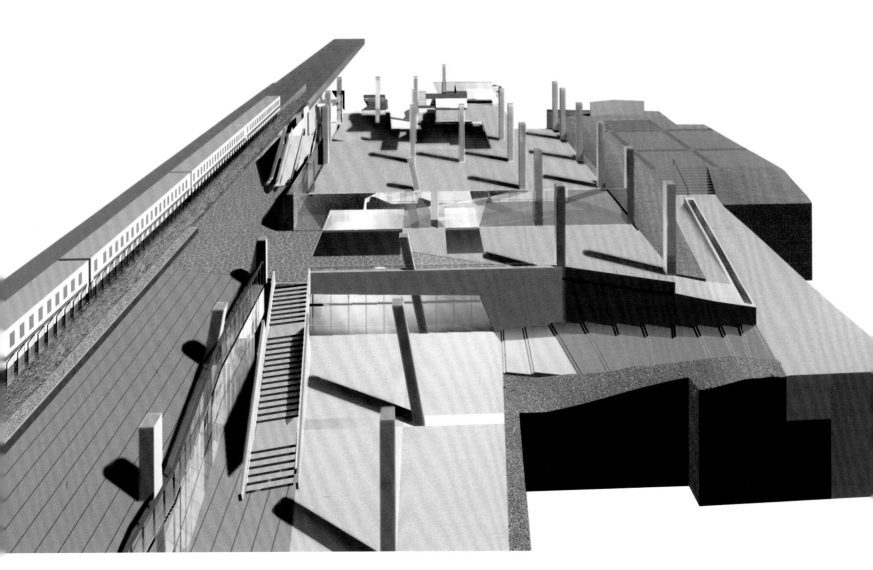

In dramatic contrast with the single-purpose nature of a traditional American development project, the brief for this mixed-use redevelopment project combined a beauty school, a public plaza, a 2,000 seat theater, an art gallery, and a residential tower in the Shinjuku district of Tokyo. Rising from the constricted parcel on which the Yamano "Gakuen" was gradually assembled in the 1950s by its gifted founder, Aiko Yamano, the school offers courses ranging from kimono wrapping and the tea ceremony to practical training for a professional career in beauty. In their design for the Yamano school, **Hodgetts+Fung** sought to craft a building that would be a thoroughly contemporary expression of globalism, while still referring to the uniquely Japanese cultural model that forms the basis of the Yamano aesthetic.

Bounded on one side by a commuter rail line, compressed by a chance mix of small residences, and entered from a narrow eight-meter right-of-way, the site itself challenged the conventions of high-rise design. In response to Tokyo's progressive solar access codes, **Hodgetts+Fung**'s design was "carved" into three separate masses, the orientation and configuration of each determined by its relationship to the surrounding cityscape and permitted envelope and floor areas.

The design reflects **Hodgetts+Fung**'s pursuit of architecture responsive to the imperatives of place, culture, and even gender. Their approach to the Yamano complex was consciously sensuous rather that technological, yielding an ensemble of forms that departed the idiom characteristic of modern high-rise projects in favor of forms that are softly folded in nature—hinting at the construction of a classical kimono. From the narrow street frontage, the ascending diagonal of the facade carries the eye upward. At the entrance, the facade "folds" to reveal a second surface, which then unveils a slender, golden tower reflecting the puzzle pieces of this dense and complex site.

 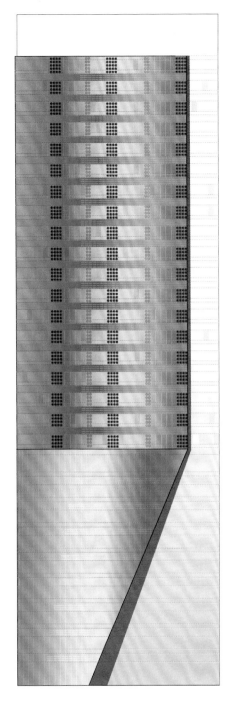 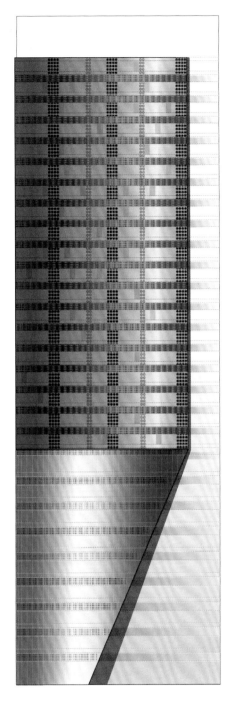

与单一用途的传统美国建造项目形成极大的对比的是,这个位于东京新宿区的多重用途的再开发项目包括一个美容学校、一个公共广场、一个可容纳2 000人的剧院、一个艺术画廊和一个住宅塔楼。山野学校局促的用地是20世纪50年代由发起人山野爱子逐渐集资起来的,这个学校提供的课程从折叠和服和茶道到一个职业美容师的训练。霍德盖茨＋冯在设计这个项目过程中,试图创造一个既表现出完全国际化的特点,又能保留独特的日本文化模式,而后者是山野美学的基本出发点。

一边是上下班火车道,一个小型的居民区占据了部分用地,加上入口是一个8米宽的通道,这个用地本身已经足够挑战常规的高层住宅设计了。为了适应东京日益增长的日照间距规范,霍德盖茨＋冯的设计像是切出的三个独立体块,体块的朝向和形状都是由它们和周围城市的关系和所允许的建筑容积和面积所决定的。

这个设计反映出了霍德盖茨＋冯追求建筑设计应当考虑已知的场地、文化、甚至性别要求。他们设计山野综合体的倾向是很感性的而不是过度技术化,产生的形状是从现代的高层建筑形态的语汇特点而来,并且轻巧地重合到自然中去——隐喻了传统和服的构造。从狭窄的街面上,立面上升的对角线把人的视线引导向上。在入口处,立面"折叠"显示出另一个表面,因此而展示出一个更纤细的、金色的塔楼,反射着这个密集的综合体和复杂的地段上的各个片断。

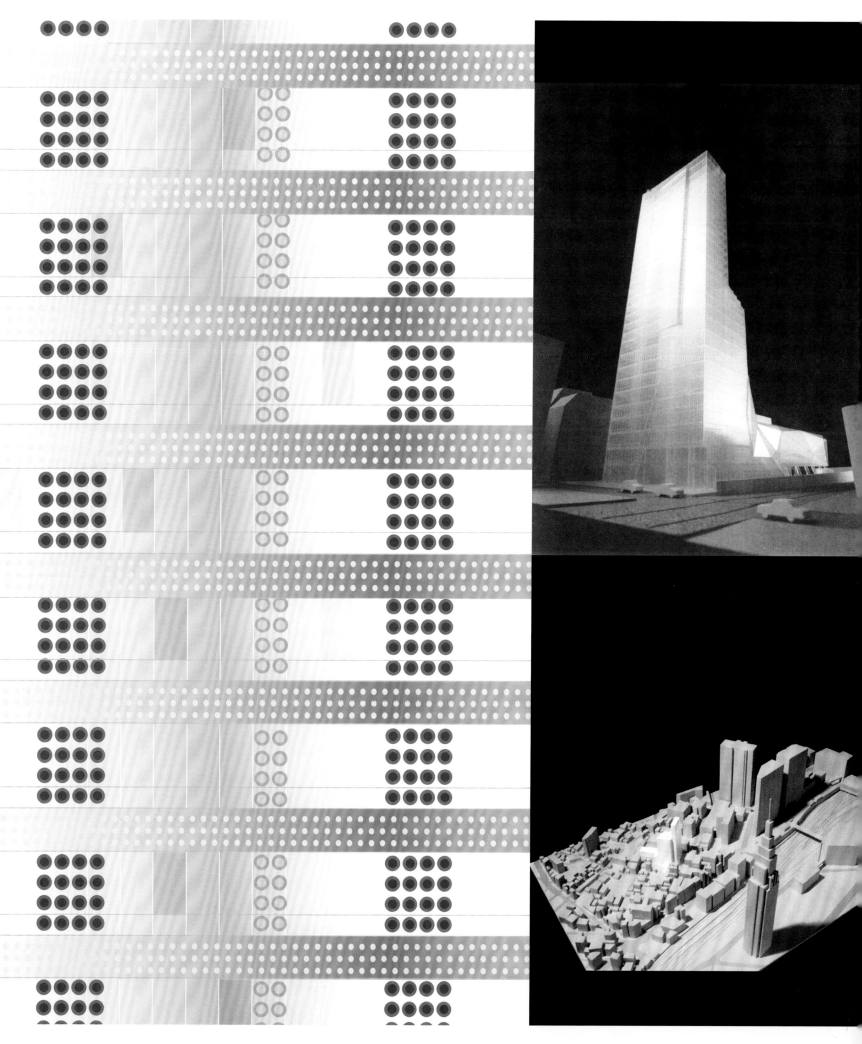

Biographies

CRAIG HODGETTS

(Born 1937)

1984–present

Hodgetts + Fung Design and Architecture, Harmonica Inc., Culver City, California, Principal and Co-Founder

1968–1983

Studio Works, New York City & Venice, California, Principal and Co-Founder

1967

Conklin and Rossant, Architects, New York City, Designer

1967–1968

James Stirling, Architect, London & New York, Designer

Teaching and Academic Administration

2000

Saarinen Visiting Professor of Architectural Design
School of Architecture, Yale University, New Haven, Connecticut

1996

Herbert Baumer Distinguished Visiting Professor
Austin E. Knowlton School of Architecture, Ohio State University, Columbus

1995

Saarinen Visiting Professor of Architectural Design
School of Architecture, Yale University, New Haven, Cunnecticut

1994–present

Professor
School of the Arts and Architecture, University of California, Los Angeles

1990–1994

Professor
Graduate School of Architecture, University of California, San Diego

1982–1990

Adjunct Professor
Graduate School of Fine Arts, Department of Architecture, University of Pennsylvania, Philadelphia

1972–1982

Graduate Studio design faculty
School of Architecture and Urban Planning, University of California, Los Angeles

1969–1972

Associate Dean
School of Design, California Institute of the Arts, Valencia

Professional Affiliations

1994–2000

Architectural Foundation of Los Angeles

1994–1995

Urban Design Advisory Coalition

1992–1998

Architecture & Design Council, Museum of Contemporary Art, Los Angeles

1991–present

Advisory Board, Los Angeles Forum for Architecture and Urban Design

1975

Second Federal Design Assembly Task Force

1975

California Council on Design Education

Education

1966

MArch, Yale University, New Haven, Connecticut

1960

BA, Oberlin College, Ohio

HSIN-MING FUNG

(Born 1953)

1984–present

Hodgetts + Fung Design and Architecture, Harmonica Inc., Culver City, California, Principal and Co-Founder

1980–1984

Charles Kober & Associates, Los Angeles, Designer

Teaching and Academic Administration

2002–present

Director, Graduate Programs and Design Faculty
Southern California Institute of Architecture (SCI-Arc), Los Angeles

2000

Saarinen Visiting Professor of Architectural Design
School of Architecture, Yale University, New Haven, Connecticut

1996

Herbert Baumer Distinguished Visiting Professor
Austin E. Knowlton School of Architecture, Ohio State University, Columbus

1995

Saarinen Visiting Professor of Architectural Design
School of Architecture, Yale University, New Haven, Connecticut

1985–2003

Graduate Program Coordinator and Design Faculty
College of Environmental Design, California State Polytechnic University, Pomona

Professional Affiliations

2004

The Trusteeship, an Affiliate of the International Women's Forum

Los Angeles Institute for the Humanities, University of Southern California,

2001–2003

Council Member, National Endowment for the Arts, Washington, D.C.

2002

Board of Directors, Westside Urban Forum, Los Angeles

1994–2000

Architectural Foundation of Los Angeles

1994–1997

President, Los Angeles Forum for Architecture and Urban Design

1994–1995

Women's Transportation Coalition, Los Angeles

Urban Design Advisory, Los Angeles

1993–present

Association for Women in Architecture

1992–present

Society of Fellows, American Academy in Rome

1992–1998

Architecture & Design Council, Museum of Contemporary Art, Los Angeles

1991–present

Advisory Board, Los Angeles Forum for Architecture and Urban Design

Education

1980

MArch, University of California, Los Angeles

1977

BA, California State University, Dominguez Hills

Selected Bibliography

Selected Publications

Coco Brown, et alia, *American Dream: The Houses at Sagaponac. Modern Living in the Hamptons*. New York: Rizzoli, 2003.

James Steele. *Architecture Today*. London: Phaidon, 1998.

John Thackara. *Design Masters: Makers of the New Culture*. Amsterdam: Weidenfeld and Nicholson, 1998.

Kurt W. Forster, "The Supercalifragilistic Architecture of Hodgetts+Fung." In *Hodgetts+Fung: Scenarios and Spaces*. New York: Rizzoli, 1997.

Uta Kreikenbohm. "Sunpower—Experience the Strength of the Sun." In *Frank O Gehry: Das Energie-Forum-Innovation in Bad Oeynhausen*. Manfred Ragati and Uta Kreikenbohm, eds. Bielefeld: Kerber Verlag, 1996.

Philip Jodidio. *Contemporary California Architects*. Cologne: Taschen Verlag, 1995.

Kurt W. Forster, "Beauty (and the beast) in the Parlor: Hodgetts+Fung in the (C)age of the Media Culture." In *A+U / Architecture + Urbanism Hodgetts+Fung* (1991): 68–128.

Philip Langdon. *Orange Roofs, Golden Arches: The Architecture of American Chain Restaurants*. New York: Knopf, 1986.

Selected Exhibition Catalogues

At the End of the Century: One Hundred Years of Architecture. Richard Koshalek and Elizabeth A. T. Smith, eds. Los Angeles: The Museum of Contemporary Art and New York: Harry N. Abrams, Inc., 1998.

Fabrications. Pat Morton, Aaron Betsky, Terence Riley, and Mark Robbins, eds. New York: Museum of Modern Art; San Francisco: San Francisco Museum of Modern Art; Columbus, OH: Wexner Center for the Arts, 1998.

Identity and Difference: The Imageries of Difference. The Triennale in the City. Pietro Derossi, ed. Milan Triennale. XIX Esposizione Internazionale. Milan: Electa, 1996.

Angels & Franciscans: Innovative Architecture from Los Angeles and San Francisco. Bill Lacy and Susan de Menil, eds. New York: Rizzoli, 1992.

Visionary San Francisco. Paolo Polledri, ed. San Francisco: San Francisco Museum of Modern Art & Munich: Prestel-Verlag, 1990.

The New Urban Landscape. Richard Martin, ed. New York: Olympic & York Companies & Drenttell Doyle Partners, 1990.

Blueprints for Modern Living: History and Legacy of the Case Study Houses. Esther McCoy, Thomas S. Hines, Helen Searing, Kevin Starr, Elizabeth A. T. Smith, Thomas Hine, Reyner Banham, and Dolores Hayden. Los Angeles: Museum of Contemporary Art; Cambridge, Mass. and London: The MIT Press, 1989.

Selected Articles

Sara Hart, "The Art and Science of Peace and Quiet." In *Architectural Record* (Febuary 2005): 143–146.

Deborah Snoonian, "Hollywood Bowl, California." In *Architectural Record* 193/1 (January 2005): 152–154.

"Hodgetts+Fung vote for Democracy C*H*A*D* Defines Current State of Voting." In *ContractMagazine.com* (October 2004).

Linda Burnett, "Twist and Bend." In *ContractMagazine.com* (October 2004).

"New Hollywood Bowl Design Recalls Former Glamour." In *AIArchitect* (September 2004).

Allison Milionis, "Hollywood Bowl is Fine-tuned and Reimagined." In *Architectural Record* (August 2004).

Jesse Brink, "The Hollywood Bowl." In *LA Architect* (July/August 2004): 20.

Mark Swed, "A Sonic Work in Progress." In *The Los Angeles Times* (24 July 2004), Calendar Section, E1.

"Quirky Gallery Opens Inside Disney Hall." In *LA Downtown News* (19 July 2004), 13.

Ryan Aday, "A New Tune for the Hollywood Bowl," In *Welding Journal* (July 2004).

"Hollywood Bowl: Music Under the Stars." Interview with Craig Hodgetts and Ming Fung for KLCS-TV, Los Angeles, July 2004.

Mark Swed, "Sonically, A Shell of Its Former Self." In *The Los Angeles Times* (28 June 2004), Calendar Section, E1 and E4.

Louise Roug, "Gehry's Globes Hanging Around After All." In *The Los Angeles Times* (26 June 2004), Calendar Section, B4.

Kate Berry, "Bowl Role." Interview with Hodgetts + Fung, in *Los Angeles Business Journal* (21 June 2004).

Nicolai Ouroussoff, "Not Much of a Shell Shock." In *The Los Angeles Times* (13 June 2004), Calendar Section.

Sara Lin and Doug Smith, "New Shell Takes a Bow at the Bowl. In *The Los Angeles Times* (10 June 2004), Section A, 1.

Greg Goldin, "Al Fresco: The New Old Hollywood Bowl." In *LA Weekly* (4 June 2004), 40.

Rebecca Winzenried, "Surround Sound." In *Symphony* (May/June 2004): 23–29.

Sharon Waxman, "Refurbishing a Fabled Bowl." In *The International Herald Tribune* (14 May 2004), International Traveler Section, 9.

Sharon Waxman, "A Fabled Bowl, Updated." In *The New York Times* (9 May 2004), Travel Section, 12 and 18.

Anne Midgette, "Filling Hollywood's New Bowl." In *The New York Times* (2 May 2004), Arts & Leisure Section, 33 and 39.

Chris Pasles, "A Season of Showing off the Bowl." In *The Los Angeles Times* (12 March 2004), Calendar Section, E2.

"Reading the Crowd." In *Metropolis* (December 2003): 60.

Suzanne Stephens, "Hodgetts+Fung Reworks LA Modernist Vocabulary Resulting in a Sleek and Linear Design for the Sylmar Library." In *Architectural Record* (November 2003): 153–157.

"Breaking Open the Box." In *Architecture Magazine* (April 2002): 104–111.

Nicolai Ouroussoff, "The Once and Future Hollywood Bowl: Architecture." In *The Los Angeles Times* (9 September 2000), Calendar Section, F1.

Mark Swed, "The Once and Future Hollywood Bowl: Music." In *The Los Angeles Times* (9 September 2000), Calendar Section, F1.

Nora Zamichow, "Hollywood Bowl May Shed Famous Shell." In *The Los Angeles Times* (26 August 2000), Section A, 1 and 19.

Nicolai Ouroussoff, "Architecture Under Glass." In *The Los Angeles Times* (9 July 2000), Calendar Section, 7 and 70.

"New Case Study Housing: MoCA Housing Competition, Franklin/La Brea, Hollywood, 1988." In *Global Architecture: Houses* 29 (July 2000): 16–21.

Kate Hensler Fogarty, "Best Entertainment Venue." In *Interiors Magazine 21st Annual Interiors Awards* (January 2000): 64–67.

Herbert Muschamp, "Molding a Plywood Utopia." In *The New York Times* (15 October 1999), Weekend Section B, 31 and 35.

Thomas Hine, "Exhibitions: Charles and Ray Eames." In *Architectural Record* (October 1999): 43–44.

"Briefs/Design: Three Design Firms Win AIA/CC Honor Awards." In *California Real Estate Journal* (June 1999): 32.

John Calhoun, "Unearthing a Rare Egyptian Artifact." In *Entertainment Design* (May 1999): 34–37.

Aaron Betsky, "Egyptian Revival." In *Architecture Magazine* (March 1999): 110–115.

Michael Webb, "American Cinemathèque at the Egyptian Theatre." In *Interiors* 158 (February 1999): 48–53.

"Conversion of the Egyptian Theatre into an American Cinemathèque, Hollywood." Photographs by Tom Bonner. In *Domus* 816 (June 1999): 11–19.

Clifford A. Pearson, "Reinventing the Mall: Universal Studios Experience, Beijing, China." In *Architectural Record* (October 1999): 151, 162–163.

Nicolai Ouroussoff, "Urban Visionaries Coming into Focus." In *The Los Angeles Times* (27 December 1998), Calendar Section, 64–65.

Robert W. Welkes, "Ancient Egyptian's Reincarnation." In *The Los Angeles Times* (3 December 1998), Calendar Section, 6–15.

Nicolai Ouroussoff, "Architecture Review." In *The Los Angeles Times* (3 December 1998), Calendar Section, 6–15.

Nicolai Ouroussoff, "An Oasis in the Making." In *The Los Angeles Times* (13 December 1998), Calendar Section, 63 and 66.

Robert Koehler, "Ready for its Close-up." In *Daily Variety* (3 December 1998), Section A, 1–4.

David Dillon, "Exhibitions: Equal Partners." In *Architectural Record* (December 1998): 28–29.

Michael J. O'Connor, "An Orchard for Artists: Villa Montalvo, Saratoga, California." In *Architecture Magazine* (December 1998): 42–45.

Sharon Waxman, "Return of the Egyptian." In *The Washington Post* (27 November 1998), Section D, 1 and 7.

Nicolai Ouroussoff, "Putting Up Walls that Break Down Barriers." In *The Los Angeles Times* (15 November 1998), Calendar Section, 65–66.

Martin Filler, "Constructing Couples." In *House Beautiful* (November 1998): 114–117, and 122.

Jennifer Minasian, "The Egyptian Theater Take Two." In *LA Architect* (July/August 1998): 28–31.

Dennis Drabelle, "Return of the Egyptian." In *Preservation News* (July/August 1998): 14–15.

Todd S. Purdum, "Hollywood is Digging Out to Restore its Heritage." In *The New York Times* (9 June 1998), Arts Section.

Carla Rivera, "Bringing Back the Past." In *The Los Angeles Times* (10 April 1998), Metro Section, Part B, 1.

Catherine Slessor, "Eamesian Invention." In *Architecture Magazine* (November 1997): 43.

Cathy Lang Ho, "Inventing Temporary Beauty." In *Metropolis* (May 1997): 76–81, and 131–133.

Nicolai Ouroussoff, "Splendor on the Boulevard." In *The Los Angeles Times* (19 April 1997), Section Γ, 1.

"Sun Power: No More Daisy," Karrie Jacobs, moderator. In *International Design Magazine* (July/August 1996): 144–145.

Klaus Schmidt-Lorenz, "Ökologische Dekonstruktion." In *Design Report* (November 1995): 64–67.

Peter Hyatt, "Advanced Learning Curves." In *Steel Profile* (September 1995): 18–25.

"Contemporary Vernacular Rehab: The Los Angeles Craft and Folk Art Museum Gets a Face-Lift." In *Design Journal* (June 1995): 16–17.

Daniel Gregory, "Stylish isles." In *Sunset* (April 1995): 140.

Ruggero Borghi, "Verde Giallo Viola." In *Ville Giardini* (March 1995): 10–15.

Abby Bussel,"The (Social) Art of Architecture." In *Progressive Architecture* (January 1995): 43–46.

Philip Jodidio, "Architects Californiens." In *Connaissance des Arts* (October 1994): 100–139.

Paola Antonelli, "Economy of Thought, Economy of Design." In *Abitare* (May 1994): 243–249.

Fulvio Irace, "The Building of an Architectural Identity." In *Abitare* (May 1994): 212–217.

Frank F. Drewes, "Bibliothek UCLA in Los Angeles, USA" In *Deutsche Bauzeitschrift* (April 1994): 49–54.

Amy Milshtein, "Hot Towell." In *Contract Design* (March 1994): 60–63.

Reinhard Renger, "Regie Für Buntd Würfel." In *Ambiente* (January/Febuary 1994): 96–101.

Aaron Betsky, "The *ID* Forty: Contemporary Solutions." In *International Design Magazine* (January/February 1994): 49.

David Leclerc, "Bibliothèque provisoire Towell." In *L'Architecture d'Aujourd'hui* (December 1993): 92–95.

Aaron Betsky, "Casting Castle." In *Architectural Record* (October 1993): 92–95.

Herbert Muschamp, "A Bright Balloon of a Building Soars at UCLA." In *The New York Times* (8 August 1993), 29.

Abby Bussel, "Now You See It." In *Progressive Architecture* (June 1993): 104–107.

Barbara Lamprecht, "Bibliotheca Temporaria." In *The Architectural Review* (June 1993): 38–43.

Kurt Forster, "Panem et Circenses." In *A+U / Architecture and Urbanism* (May 1993): 9–31.

Undine Pröhl, "Das Ende Steht Nahe Behor." In *Hauser* (May 1993): 82–85.

Aaron Betsky, "Under the Big Top." In *Architectural Record* (March 1993): 94–101, and cover.

Aaron Betsky, "Biggest Show on Earth." In *International Design Magazine* (January 1993): 24.

Dennis Puzey, "House of Many Colors." In *House and Garden* (September 1992): 94–97.

Vernon Mays, "Carving a niche for the 90's: Entertainment Design." In *Architecture Magazine* (May 1992): 96–97.

Michael Webb, "Hollywood Blockbuster." In *Belle Design and Decoration* (February/March 1992): 60–67.

Francesca Garcia-Marques, "Arts Park, Los Angeles." In *L'ARCA* (September 1991): 40–45.

Horst Rasch, "Paradis Vogel à la Hollywood." In *Hauser* (May 1991): 14–31.

Aaron Betsky, "The New Colors of Modernism." In *Metropolitan Home* (April 1991): 150–154.

"Al Borde del Abismo, Residencia Viso, Hollywood." In *Monografías de Arquitectura y Vivienda* 32 (1991): 34–37.

"Dreamscape, Reality and Afterthoughts." In *A Quarterly Journal of Environmental Design 7/2* (Winter 1990): 12.

Michael Webb, "A Loft in the Hollywood Hills." In *LA Style Magazine* (November 1990): 184–187.

Philip Arcidi, "Projects." In *Progressive Architecture* (September/November 1990): 143–146.

Aaron Betsky, "Change in Scene." In *Architectural Record. Record Interiors* (September 1990): 104–109.

"Blueprints for Modern Living." In *Annals of the Architecture Association* 20 (Autumn 1990): 77–82.

Nancy Scott, "San Francisco Future." In *Metropolitan Home* (July 1990): 32.

Gallagher, Larry, "The Shape of Things To Come, Brave New City." In *San Francisco Magazine* (June 1990): 70–77.

"Hodgetts+Fung Design Associates: Viso Residence and Thames Residence." In *Global Architecture: Houses* 28 (March 1990): 96–99.

Douglas Suisman, "Utopia in the Suburbs." In *Art in America* (March 1990): 184–193.

Herbert Muschamp. "Craig Hodgetts and Ming Fung—You Send Me." In *Terrazzo 4* (1990): 93–108.

Raymond Ryan, "At Home with the Future." In *Blueprint 63* (December 1989/January 1990): 49–50.

Paola Antonelli, "Storia ed Eredità Delle Case Study Houses." In *Domus* (December 1989): 10–11.

Aaron Betsky, "Steel Chic and Stucco Dreams at the LA Lab." In *Metropolitan Home* (August 1989): 75–77.

"Beautification Takes Command." In *Bauwelt* 80/16–17 (April 1989): 770–771.

Leon Whiteson, "P/A Portfolio: Housing for the Future." In *Progressive Architecture* (October 1988): 96.

Steven Holt and Michael McDonough, "Apocalypse Now: The New L.A." In *Metropolitan Home* (July 1988): 24.

Steven Holt, "Creative Trends, World Pulse, New York." In *Axis* (Summer 1988): 23.

Pilar Viladas and Susan Doubilet, "U. C. Builds." In *Progressive Architecture* (May 1988): 85–93.

Aaron Betsky, "The Ephemerality of a Cinematic Architecture." In *Architectural Review* (December 1987): 59–60.

Aaron Betsky, "Villa Linda Flora: The Emerging Generation, U.S.A." In *Global Architecture: Houses* 2 (1987): 22–27.

"International Architectural Competition for '88 Seoul Olympic Village." In *Magazine of Architectural Culture* [Korea] (July 1985): 38–41.

Susan Doubilet, "Cookie Express, Architectural Design Citation." In *Progressive Architecture* (January 1985): 124–125.

Selected Projects

In Progress

YAMANO GAKUEN DEVELOPMENT PROJECT, Yamano Gakuen Beauty School, Tokyo, Japan (to be completed 2007), project team: Tetsuya Fukui, Azusa Kotsu and Chiaya Yasunori, executive architect: Rui Sekkeishitsu Co., Ltd., project management: Index Consulting, contractor: Taisei Corporation

2004

WORLD OF ECOLOGY, California Science Center, Los Angeles, (to be completed 2008), project team: Silvia Song, Curran Starkey, Katherine Harvey and Chihsuan Tsai collaborating exhibit designer: Science Museum of Minnesota

2003

PARQUE DE LA GAVIA, City of Madrid, Spain (invited competition), project team: Ron Calvo, Neil Silberstein, Kate Harvey, Curran Starkey, Michael Tadros, Asuza Kotsu and Peter Mayor, consultant: Edina Weinstein (arborist)

2002

GERSHWIN GALLERY EXHIBIT, The Library of Congress, Los Angeles, project team: Neil Silberstein (leader) and Kate Harvey

THE CALIFORNIA ENDOWMENT, Los Angeles, (invited competition), project team: Michael Knopoff, Neil Silberstein, Matias Creimer, Darren Morley, Sasha Monge and Gabrielle Schlesinger, consultant: Mia Lehrer + Associates (landscape)

TEMPE VISUAL AND PERFORMING ARTS CENTER, City of Tempe, Arizona, (invited competition), project team: Eric Holmquist, Thierry Garzotto, Neil Silberstein, Rafael Rosas, Matias Creimer, Clifford Takara and Rebecca Grant, in collaboration with van Dijk Pace Westlake and Hargrave Associates (landscape)

2001

VIRTUAL HOUSE, The Brown Companies, Sagaponac, Long Island (to be completed 2008), project team: Neil Silberstein (leader) and Thierry Garzotto

HYDE PARK BRANCH, Los Angeles Public Library, Los Angeles, project team: Greg Stutheit, David Wick, Ronald Calvo and Greg Kochnowski, consultants: Englekirk and Sabol, Inc. (structural), The Sullivan Partnership (mechanical), Patrick Byrne & Associates (electrical), Katherine Spitz & Associates (landscape) and Robin Strayhorne (artist)

2000

SINCLAIRE PAVILION, Art Center College of Design, Pasadena, California, project team: Michael Knopoff, David Wick, Greg Stutheit and Asuza Kotsu, consultant: William Koh Associates (structural)

WORLD SAVINGS AND LOAN, World Savings and Loan Bank, Alhambra, California, project team: Matias Creimer, Eric Holmquist, Andrew Lindley, Denise Zacki and Richard Bonneville, collaborating consultants: William Koh Associates (structural), KPFF Consulting Engineers (mechanical and civil)

SYLMAR BRANCH, Los Angeles Public Library, Los Angeles, project team: Ronald Calvo, Greg Stutheit, Denise Zacki, Richard Bonneville and Birgit Bastiaan, consultants: Englekirk and Sabol, Inc.(structural), The Sullivan Partnership (mechanical), Patrick Byrne & Associates (electrical), Katherine Spitz & Associates (landscape) and Barbara Strasen (artist)

1999

HOLLYWOOD BOWL AMPHITHEATER, Los Angeles Philharmonic Association, Hollywood, project team: Eric Holmquist (project architect), Denise Zacky, Martha de Plazaola, Kevin Owens and Richard Bonneville, executive architect: Gruen & Associates, consultants: Ove Arups & Partners (structural, electrical and mechanical), Fisher Dachs Associates, Inc. (lighting), Jaffe Holden Acoustics (acoustisc) and MHI Miyomoto (structural)

BROOKLYN PARK AMPHITHEATRE, Minnesota Orchestral Association, Minneapolis, (project), project team: Eric Holmquist (leader) and David Wick executive architect: Hammel, Green and Abramson, Inc., consultants: BBA Architectural Planners (theatre planning), Jaffe Holden Acoustics, Inc. (acoustics) and Oslund & Associates (landscape)

1997

UNIVERSAL EXPERIENCE, MCA/Universal, Beijing, China, project team: Martha de Plazaola, Thierry Garzotto, Denise Zacki and Janice Shimizu, associate architect: House & Robertson, project management: Welbro Management, Inc., consultants: Ove Arup & Partners (structural), Joe Kaplan Lighting (lighting), Edwards Technologies, Inc. (media systems), Olio (wrap design) and David Riley Associates (signage and logo)

ZKT WAVE POWER, Elektrizitatswerk Minden-Ravensberg, Energie Forum-Innovation, Bad Oeynhausen, Germany, project team: Michael Knopoff (leader) and John Martin, consultant: Ralph Hudson (electro-mechanical engineering), Ironwood (fabricator)

MICROSOFT PAVILION ELECTRONIC ENTERTAINMENT EXPO, Microsoft Corporation, Atlanta, project team: Selwyn Ting, Clover Lee, Denise Zacki, Robert Kerr and Alexander Desbulleux, consultant: Occidental Studios (lighting), George & Golberg (fabricator)

1996

MICROSOFT PAVILION ELECTRONIC ENTERTAINMENT EXPO, Microsoft Corporation, Los Angeles, project team: Michael Swischuk, Selwyn Ting, Shawn Sullivan, Mark Kruger, Ken Mattiuz and Sally Painter, consultant: ROYGBIV (lighting), George & Goldberg (fabricator)

1995

AMERICAN CINEMATHEQUE, American Cinematheque, Hollywood, project team: Eric Holmquist, Andrew Lindley, Michael Swischuk, Stephen Billings, Clover Lee, Janice Shimizu. Christine Cho and Alberto Reano, consultants: Englekirk and Sabol, Inc. (structural), Boston Light & Sound (projection & sound), Mckay Conant Brook (acoustics), Historic Resources Group (historic preservation)

1994

LA CITTA PULPA, XIX Milan Triennale, project team:Selwyn Ting, Michael Swischuk, Ken Mattiuz and Clover Lee

THE WORLD OF CHARLES AND RAY EAMES, The Library of Congress and Vitra Design Museum, (traveled), project team: Michael Knopoff, Yanan Par, and Douglas Pierson, consultant: Eames Office (media)

MULLIN SCULPTURE STUDIO, Occidental College, Eagle Rock, project team: Yanan Par, David Campbell, in collaboration with Levin & Associates Architects, consultants: Englekirk and Sabol, Inc. (structural), Babsco Enterprises (electrical), The Sullivan Partnership (mechanical)

ZKT SUN POWER, Elektrizitatswerk Minden-Ravensberg, Energie Forum-Innovation, Bad Oeynhausen, project team: Ken Mattiuz and Stevens Wilson, consultants: Jon Biondo (electronics), Ralph Hudson (electro-mechanical engineering), Blake Hodgetts (music composition), Ironwood (fabricator)

1991

TOWELL LIBRARY, University of California at Los Angeles, Project team: Robert Flock (leader) and Peter Noble, consultants: Robert Englekirk Engineering, Inc. (structural), Patrick Byrne & Associates (electrical), The Sullivan Partnership (mechanical), A. C. Martin & Associates (civil) and Rubb Inc. (shell)

Selected Awards

2005

Merit Award, AIA California Council
Los Angeles Public Library, Sylmar Branch

Merit Award, AIA California Council
Sinclaire Pavilion

Civic Award of Excellence, LA Business Council
Hollywood Bowl

2004

Honor Award, AIA Pasadena Foothill Chapter
Sinclaire Pavilion

Civic Award of Excellence, LA Business Council
Los Angeles Public Library, Sylmar Branch

Interiors Award of Excellence, LA Business Council
Library of Congress/Ira Gershwin Gallery

Special Award, Association of Environmental Professionals
Hollywood Bowl Design Guidelines

2003

Outstanding Public/Private Sector Civil Engineering Project
LA Branch, American Society of Civil Engineers
Hollywood Bowl

Merit Award, AIA LA Chapter
Parque de la Gavia

2002

Merit Award, AIA LA Chapter
Sinclaire Pavilion

Design Distinction Award, 48th Annual *ID Magazine* Design Review
Sinclaire Pavilion

Institutional Award of Excellence, LA Business Council
Sinclaire Pavilion

2001

Architectural Design Excellence Award, City of LA Cultural Affairs Commission
Los Angeles Public Library, Sylmar Branch

2000

City of LA Cultural Heritage Commission Award
American Cinemathèque at the Egyptian Theatre

Honor Award, National Trust for Historic Preservation
American Cinemathèque at the Egyptian Theatre

21st Annual *Interiors Magazine* Entertainment Venue Award
American Cinemathèque at the Egyptian Theatre

Entertainment Venue Award, Westside Urban Forum
American Cinemathèque at the Egyptian Theatre

Charlie Award, Hollywood Arts Council
American Cinemathèque at the Egyptian Theatre

Preservation Award, LA Conservancy
American Cinemathèque at the Egyptian Theatre

Honor Award, AIA California Council
American Cinemathèque at the Egyptian Theatre

Beautification Award, LA Business Council
American Cinemathèque at the Egyptian Theatre

1997

Design Distinction Award, 43rd Annual *ID Magazine* Design Review
Microsoft E3 '96

Best of Category, Environments, 42nd Annual *ID Magazine* Design Review
Sun Power: No More Daisies

1993

Lumen Award, International Illuminating Engineers Society
Towell Library

Merit Award, AIA California Council
Towell Library

Merit Award, AIA Los Angeles Chapter
Viso Residence

Award of Excellence Award, AIA/American Library Association
Towell Library

1992

Creative Business Award, West Hollywood Chamber of Commerce
Click and Flick Agency

1990

Record Interiors Award, *Architectural Record*
Hemdale Office Facility

1980

Special Citation, AIA Los Angeles Chapter
Blueprints for Modern Living

1984

Citation, *Progressive Architecture*
Cookie Express

1983

Citation, *Progressive Architecture*
Venice Interarts Center

1978

First Design Award, *Progressive Architecture*
Southside Settlement

1972

First Design Award, *Progressive Architecture*
00:00 A Mobile Theater

Citation, *Progressive Architecture*
Southside Settlement

1969

First Design Award, *Progressive Architecture*
Decitrun 636

Firm Awards and Prizes

2005

First Award, Menlo Atherton HS Performance Arts Center, California

2004

AIA Educator of the Year Award, Los Angeles Chapter (Craig Hodgetts)

1997, 1999

Architectural Foundation of Los Angeles

1996

Chrysler Award for Innovation in Design

1995

20 Stars of Design, Pacific Design Center, Los Angeles

1994

Architecture Award, American Academy of Arts and Letters

1994

The I.D. Forty, *International Design Magazine*

1992

Graham Foundation Grant (Hsinming Fung)

1991

NEA / Rome Prize Advanced Fellowship in Design Art (Hsinming Fung)

1989

First Award, Arts Park Los Angeles, Invited International Competition

1987

Award of Merit, West Hollywood Civic Center Competition

1985

Excellent Entry, International Competition, Olympic Village, Seoul

1984

First Award, Urban Design Competition, Little Tokyo, Los Angeles

Acknowledgements

Projects often come to us either having little or no precedent, or so laden with preconditions that the design strategy must struggle to emerge. Our determination to maintain a multi-disciplinary practice in which architecture and design are fused with a deep commitment to learning and expermentation has required extraordinary support from our collaborators—clients as well as members of the studio.

Given that our office is small and our expectations great, our staff has developed the capacity to respond quickly and generously to the challenges of a truly diverse array of projects—from public projects like our Hyde Park and Sylmar branch libraries for the City of Los Angeles to the highly politicized Hollywood Bowl; from the deeply researched histories presented in exhibitions like the World of Charles and Ray Eames to the nearly instantaneous birth of the Towell Library for UCLA.

Along the way, we have worked with and learned from many extraordinary individuals on projects which reached their full potential largely due to their vision and tenacity. Rolf Felhbaum of Vitra Design Museum and Irene Chambers from the Library of Congress were trusted guides as we navigated the interpretive chasms among the many institutions that participated in the Eames exhibition.

Less vulnerable, but equally in need of a visionary owner, was the Egyptian Theater, which was transformed from a derelict husk to become a catalyst for Hollywood's redevelopment by the dedication of executive director Barbara Smith. Ron Lindsay, a builder whose passion matched our own, went to great lengths on the same project to see that our design was faithfully mirrored in the finished project.

Sometimes even a modest project, like that at Art Center, can have an outsized impact, especially relative to the pleasure we derived from working once again with Richard Koshalek, who refused compromises and enabled us to achieve our vision.

We must also thank the Yamano family for entrusting their vision to us, and giving us the opportunity to work in Tokyo. It has been an inspiring and rewarding journey, transforming our office, and challenging us to absorb the aesthetic systems of a wonderful culture.

We would not have been able to realize this book if not for the dedication of Tetsuya Fukui, Asuza Kotsu, Silvia Song, Joanna Ellis, and Danielle Butts.

We would like to especially recognize Patti Rovtar, who has faithfully supported our design priorities with her resourceful approach to our daily operations, even when we were spread too thin.

Finally, we must thank those who believe in our practice. Reed Kroloff has helped us to gain critical visibility and tangible opportunities through his unstinting editorial support. Richard King and Eddie Wang offered encouragement and advice, introducing us to a global perspective at a critical time and continuing to offer their support as mentors and friends. Denise Bratton has been a tireless and perceptive interpreter, editing this volume, and lending wisdom and intelligence to the task of making sense of our work. Once again, we are touched by the wit and insight of Kurt Forster, who always seems to know more about our idiosyncratic thought processes and out-of-the-box working methods than we do.

图书在版编目(CIP)数据

霍德盖茨+冯／蓝青主编，美国亚洲艺术与设计协作联盟(AADCU).
北京：中国建筑工业出版社，2005
（美国当代著名建筑设计师工作室报告）
ISBN 7-112-07523-8

Ⅰ.霍... Ⅱ.蓝... Ⅲ.建筑设计－作品集－美国－现代 Ⅳ.TU206
中国版本图书馆CIP数据核字(2005)第080878号

责任编辑：张建 黄居正

美国当代著名建筑设计师工作室报告
霍德盖茨+冯

美国亚洲艺术与设计协作联盟(AADCU)
蓝青 主编
*
中国建筑工业出版社 出版、发行（北京西郊百万庄）
新华书店经销
北京华联印刷有限公司印刷
*
开本：880×1230毫米 1/12 印张：13
2005年8月第一版 2005年8月第一次印刷
定价：**128.00元**
ISBN 7-112-07523-8
　　(13477)
版权所有 翻印必究
如有印装质量问题，可寄本社退换
（邮政编码 100037）
本社网址：http://www.china-abp.com.cn
网上书店：http://www.china-building.com.cn